Project Management in Construction

Project Management in Construction

Third Edition

Anthony Walker

MSc, PhD, FRICS, FHKIS, MCIOB

Professor of Surveying
University of Hong Kong

Blackwell
Science

Copyright © Anthony Walker 1984, 1989, 1996

Blackwell Science Ltd
Editorial Offices:
Osney Mead, Oxford OX2 0EL
25 John Street, London WC1N 2BL
23 Ainslie Place, Edinburgh EH3 6AJ
238 Main Street, Cambridge,
   Massachusetts, 02142, USA
54 University Street, Carlton,
   Victoria 3053, Australia

Other Editorial Offices:
Arnette Blackwell SA
   224, Boulevard Saint Germain
   75007 Paris, France

Blackwell Wissenschafts-Verlag GmbH
   Kurfürstendamm 57
   10707 Berlin, Germany

   Zehetnergasse 6
   A-1140 Wien
   Austria

First edition published by Granada Publishing 1984
Second edition published by BSP Professional Books
   1989
Third edition published by Blackwell Science 1996

Set in 10 on 12½ pt Times
by DP Photosetting, Aylesbury, Bucks
Printed and bound in Great Britain by
Hartnolls Ltd, Bodmin, Cornwall

The Blackwell Science logo is a
trade mark of Blackwell Science Ltd,
registered at the United Kingdom
Trade Marks Registry

DISTRIBUTORS

Marston Book Services Ltd
PO Box 269
Abingdon
Oxon OX14 4YN
(*Orders:* Tel: 01235 465500
            Fax: 01235 465555)

USA
Blackwell Science, Inc.
238 Main Street
Cambridge, MA 02142
(*Orders:* Tel: 800 215-1000
            617 876-7000
            Fax: 617 492-5263)

Canada
Copp Clark, Ltd
2775 Matheson Blvd East
Mississauga, Ontario
Canada, L4W 4P7
(*Orders:* Tel: 800 263-4374
            905 238-6074)

Australia
Blackwell Science Pty Ltd
54 University Street
Carlton, Victoria 3053
(*Orders:* Tel: 03 9347-0300
            Fax: 03 9349-3016)

A catalogue record for this title
is available from the British Library

ISBN 0–632–04071–8

Library of Congress
Cataloging-in-Publication Data
is available

Contents

Preface

The effective management of construction projects continues to be a challenge to project managers and their project teams. Whilst the focus of most books on construction management remains on procedures and techniques, the significance of organisation design and the need for an understanding of the project management process have become increasingly recognised. This book focuses on these issues and since it was first published much progress has been made in furthering our understanding from both theoretical and practical perspectives. Nevertheless the continuing complexity of projects and their environments means that much remains to be learned. It seems that this book has made some contribution to understanding on the basis of feedback I have received from postgraduate students and those in industry; for which I am grateful. But whilst the theoretical framework addressed in this book can be sustained, its base now needs reinforcing and there are other areas which are in need of strengthening and new topics which need adding.

To underpin the theoretical core, a new chapter on organisations generally and their relevance to the construction process is added which uses some of the material in the previous edition's chapter on systems. This new chapter draws on the work of Scott (1992) particularly in developing further understanding of the variety of approaches to analysing organisations. The systems chapter remains but room is found for a more fundamental examination of systems theory leading to a substantially revised chapter.

A new chapter on authority, power and politics is introduced. The authority structure in projects was implicit within earlier editions but is now made explicit together with a review of the significance of the literature on power and politics in organisations to projects. As in the last edition, the chapter on clients is once again strengthened as is the chapter on project teams. The cases in the case study chapter, introduced in the last edition, have been simplified and a further case study has been added which has features distinctly different from the original case studies and which is analysed from a broader perspective.

Whilst organisation structure is quite firmly the focus of the book, behavioural factors are not ignored and the postscript has been completely revised to look ahead to the likely future impact of such phenomena on research into project organisations. The remainder of the book has been generally revised and updated and much greater attention has been paid to the bibliography which will now better serve as a guide to further reading.

I am indebted to my colleagues at the University of Hong Kong, particularly Steve Rowlinson, Anita Liu and Chau Kwong Wing for our discussions and their invaluable advice. My thanks are also due to Kannex Chu and Polly Chang for their patience and skill in managing to type my most untidy draft and to my wife for her support and encouragement.

I would again like to acknowledge all those I included in the Prefaces to the First and Second Editions, particularly Alan Wilson and Arthur Britch who helped me to develop my initial ideas, and also the many other members of the construction and real estate industries and academic colleagues who in discussion have helped to shape the way I have approached the subject of this book. The process is incremental, each person adds something, I simply reflect on what they have written or said. Nevertheless, I remain solely responsible for any faults which remain in this book but hope that it contributes something to each person who reads it.

Anthony Walker
Hong Kong
1996

Chapter 1
Relevance of Organisation Structure

Introduction

The management of construction projects has been carried out since people first co-operated to erect buildings yet there is little documented knowledge of how people interact in this process. It is revealing that historical and contemporary accounts of construction work pay little attention to how people worked together and managed their activities.

Writers over the ages have concentrated upon the buildings themselves, particularly on aesthetics, the use of new materials, technological developments and the impact of buildings on their environment. How people were organised and managed has received scant attention. What has been written has tended to be about such charismatic characters of enormous ability as Brunel and Wren, and not about how they structured their organisations. Refreshing in this respect is Alan Wilson's (1976) contemporary account of how the construction of Rhuddlan Castle was managed by Master James of St. George. He completed the works using 3000 men in less than three years, and that was over 700 years ago. Perhaps we would see a significant improvement in performance if the managers of the construction process today were given titles like 'Master James'! A more recent contribution is made by Morris (1994) who documents how many contemporary projects were organised and implemented. Many are outside the construction field and include, for example, Polaris and Concorde, but these illustrations of project management in practice also make a valuable contribution to project management in construction.

The way in which the available skills are used is of paramount importance in providing what clients expect from their projects. There is little point in the construction industry developing the special skills of its members if no one is going to amalgamate them in the best manner to meet a particular client's objective.

The conventional method of organisation for construction projects, by which is meant one in which the architect or engineer is designer and manager of the process, using specialist consultants with the construction contract awarded by competitive tender after the design is substantially complete, evolved in response to pressures other than solely the needs of clients. It developed in conditions that were considerably more stable than those faced today by both the construction industry and its clients.

The complexity of the conditions within which the construction industry's clients exist makes them place increasing demands upon the industry in terms of the performance of projects (both functionally and aesthetically), the capital and running costs and the time required from conception of the project to occupation. This has come about as a result of technological developments, uncertain economic conditions, social pressures, political instability, etc. Within such conditions, clients in both private and public sectors have to increase their effectiveness to remain competitive and to satisfy their own clients who transmit the demands of a complex world to them. The construction industry has in turn to respond to demands from clients that arise from such conditions and is itself also subject to external pressures in a manner similar to that of its clients. It therefore needs to respond by mobilising the talents it possesses in a way which recognises the particular needs of individual clients. It has become increasingly recognised that it is unreasonable to suppose that the conventional way of organising construction projects is likely to be a universal solution to producing a project in today's conditions.

The complexity of clients' demands, together with the increasing complexity of building, civil and industrial engineering, and other constructional work, particularly as a result of technological developments, has over the years resulted in specialisation within the construction industry. The professions associated with construction have emerged as separate skills (e.g. architecture, quantity surveying, structural, mechanical and electrical engineering, acoustics, safety), as have the many specialist subcontractors working with the general contractor. On any project, even a small one, a large number of contributors and skills are involved. On the largest there is a vast range of skills and materials required and an enormous variety of people and equipment to mobilise. Where these projects are carried out overseas, there are many additional problems of logistics and language.

The key to the management of construction projects is therefore the way in which the contributors are organised so that their skills are used in the right manner and at the right time for the maximum benefit to the client. There is little point in the construction industry developing its skills if they are not then implemented effectively.

The way in which the industry and its skills and professions have evolved has compounded the problem of organising effectively. Specialisation has been accompanied by the creation of independent companies offering the specialisations, and the complexity of construction has led to greater interdependency between the specialisations and hence between companies. This produces a high level of differentiation within the construction process and a consequent need for strong integration of the independent companies and skills. This situation has been reinforced by professional allegiances which, in the UK and elsewhere, have been compounded by the establishment of professional institutions, which in turn have contributed to the division of the design professions among themselves and their separation from construction firms.

It is against this background that the conventional solution to project organisation has attempted to cope with increasing complexity and uncertainty. The strain has shown in recent years through the increasing use and development of alternative approaches such as design-and-build, management contracting and construction management. What these approaches have not provided, however, is a framework for designing organisations to suit the particular project and the conditions in which it has to be executed. The pressure from clients has made the professions and industry take more seriously the need for organisation design, which is fundamental to the improvement of the project management process.

It should be clear by now that this book views the most important element of project management as an organisational issue which incorporates the way in which people are organised and managed in the project management process. This is a long step from the view of project management taken by many which sees it as a collection of planning and control techniques and other management tools and decision-making techniques. The distinction is important as the use of techniques and tools, however sophisticated, will be of no avail if they are applied within inappropriate organisations seeking to achieve misguided objectives. Objectives and organisation must come first if the use of planning and control techniques are to be effective in providing the information on which management decisions can be based.

Management and organisation

Before discussing project management and particularly organisation structure, it is necessary to have a clear idea of what is meant by management and by organisation. It is hardly surprising that definitions of management have occupied authors of management literature at length when the *Shorter Oxford English Dictionary* lists ten meanings of 'to manage', ranging from 'training a horse', and 'wielding a weapon' to 'controlling the course of affairs by one's own action'. The minds of many are also conditioned by its ironical use, which the dictionary quotes as 'to be so unskilful or unlucky as to do something'. Much of the literature presupposes that the reader has a clear idea of the concepts of management and organisation. Some writers offer a dictionary-style definition, but the operational definitions offered by Cleland and King (1983) are perhaps the most useful.

An operational definition is one that identifies a number of observable criteria, which, if satisfied, indicate that what is being defined exists. Cleland and King's operational definition of *management* identifies the criteria of 'organised activity, objectives, relationships among resources, working through others and decisions'. In providing an operational definition of *organisation*, Cleland and King had to employ many of the elements used in their management definition.

Organisation and management are intrinsically interlinked concepts. The former is concerned with the 'organised activity' part of their definition of management, and their observable criteria are 'objectives, some pattern of authority and responsibility between the participants with some non-human elements involved'. Decisions, both routine and strategic, are required from management to make the organisation operate.

Although management and organisation are closely interlinked concepts, it is interesting to note that management is more frequently defined in the literature that is organisation. Yet it was said (Likert 1961) many years ago that 'how best to organise the efforts of individuals to achieve desired objectives has been one of the world's most important, difficult and controversial problems' and it still holds today. It may be said that, in industries more homogeneous than the construction industry, the distinction between management and organisation is sufficient, but an especially sharp focus on the organisation of the many diverse contributors to construction projects is necessary if the successful management of projects is to be achieved.

For the purpose of accomplishing a construction project, an organisation can be said to be the pattern of interrelationships, authority and responsibility that is established between the contributors to achieve the construction client's objectives. Management is the dynamic input that makes the organisation work. When this takes place, the organisation ceases to be static and works and adapts to meet the objectives laid down for it. Management is therefore concerned with setting, monitoring and adapting as necessary the objectives of the project organisation as transmitted by the client, and with making or advising on the decisions to be made in order to reach the client's objectives. This is achieved by working through the organisation set up for this purpose, which is particularly difficult for construction projects owing to the temporary nature of most project organisations. In many cases, members of the organisation are part-time, as they are also involved in other projects and are normally seconded from their parent company.

The contributors to the project act through the organisation that has been established to carry out their work, and they produce information that allows the managers of the project to make the decisions that will keep the process going. The effectiveness of the organisation structure is therefore fundamental to the quality both of the information on which decisions will be taken and of the decision-making process itself.

Building versus civil engineering

In many countries building and civil engineering are seen as practically separate industries. Whilst fundamentally they use the same technology, the scale and intensity of the use of different technological components varies greatly. As a result the basis of design differs and consequently consultants tend to specialise in either building or civil engineering work. Also, as public sector clients are by far the greatest procurers of civil engineering work, they are really the only clients involved regularly in both building and civil engineering work. On the other hand many contractors frequently undertake both types of work.

Whilst design and construction were traditionally not separated in civil engineering in as severe a manner as in building, nevertheless the consulting engineer acted in a way which was similar to the architect. That is, they both administered the contract and had a quasi-judicial role in disputes. In both cases the contractor was usually appointed on the basis of a competitive bid and entered into a contract with the client based on a standard form. The resulting organisation structure was therefore similar for civil engineering and buildings with high differentiation of specialist consultant contributors, contractors and subcontractors but with no formal recognition of the need for integration.

The application of organisation theory is therefore as relevant to civil engineering as it is to building. The design of both civil engineering and building project organisations will benefit from the application of the ideas arising from the issues discussed in this book. Project management has, over recent years, been taken much more seriously in both sectors demonstrating the parallel need identified by sponsors and managers of projects. Further progress will be made through a greater understanding of the basis of project management which will arise from the common view of theoretical work identified in this book even though the applications relate predominantly to building.

Definition of construction project management

General management definitions require amplifying before they can be used for defining construction project management, which can be said to be:

> 'The planning, co-ordination and control of a project from conception to completion (including commissioning) on behalf of a client requiring the identification of the client's objectives in terms of utility, function, quality, time and cost, and the establishment of relationships between resources, integrating, monitoring and controlling the contributors to the project and their output, and evaluating and selecting alternatives in pursuit of the client's satisfaction with the project outcome.'

In this context, resources is a general term, which includes materials, equipment, funds and, in particular, people. A fault with many current definitions of project management is that they do not make a specific reference to managing people to achieve a project. Although it can be implied that projects can only be achieved by working through others, nevertheless it is important that definitions make explicit reference to this fundamental aspect of project management.

The implementation of this definition could take many forms in practice, depending on the nature of the project and the circumstances in which it is carried out. However, as referred to earlier, the professions still tend to seek to achieve it through a conventional organisation structure even though many new structures have been devised. Nevertheless, no matter what organisation structure is adopted, if project management is taking place, the activities identified within the definition should be observable.

Objectives and decisions

Reference to objectives and decisions has been predominant so far and they have particular significance for construction project management. The objectives of the project management process are those of the client, and a role of project management is to ensure that the project organisation works to achieve the client's objectives. Similarly, decisions taken during the process should be taken with the sole purpose of achieving the client's objectives.

Because a large number of organisationally independent firms are usually involved in construction projects and second their staff on a part-time basis, their integration and orientation to the client's objectives are major functions of project management. Thus, objectives need to be clearly stated and the head of the project management team will have to extract them from the client, state them clearly and transmit them equally clearly to the contributors to the project.

It is important that any adaption of the objectives that may subsequently occur is treated similarly. It is natural to be greatly concerned with the original objectives, but adaptions are not always given the same attention, leading to dissatisfaction with the completed project on the part of the client.

As contributors will normally be involved in a number of projects at the same time, conflicting demands upon their time and attention are always likely to occur within contributors' firms. The project organisation must be designed and managed to resolve such conflict in the interest of its client so that it does not detract from the achievement of the client's objectives.

The project management process and the project manager

The use of the title 'project manager' in the construction industry has deflected attention away from consideration of the process of project management. It is necessary at this point to distinguish between the title and the process. A common reaction seems to be that if there is someone called a project manager, then all project management problems will instantly be solved. But the project management process will take place irrespective of the titles of the people in the process. The industry needs to be concerned with identifying and studying the process of managing construction projects and with structuring its organisations and implementing techniques and procedures that make the process more effective. It may well be that the designation of a suitable individual with the title of project manager will assist in this, but it is not likely to be an instant and universal solution.

The approach should be to identify the process to be undertaken for the achievement of the specific objectives of the client, the conditions in which it is to be carried out and the people available for the project. As a result of this analysis, the organisation structure should be designed to suit the particular project. The nature of the project should establish the roles of the contributors and ascertain whether or not a role emerges that requires the title of 'project manager' as a reflection of the project's needs. Such an approach would focus attention upon the process of project management with the result that effort should be put into making it more effective rather than into a preoccupation with titles.

The title 'project manager' should have a reserved meaning in the construction industry. Projects are executed for clients and as the title means managing the project as a whole, then it should refer to managing the project for the client: that is, the specific and unwavering objective of the project manager must be achievement of the client's objectives. The project manager will therefore seek to resolve conflict in the process in the interests of the client. This implies that the project manager should be a member of the client's organisation.

One step removed from this, and more practically, the project manager could be acting as a professional consultant without an entrepreneurial interest in the project. Even in this latter case it is possible to conceive of a situation in which project managers might have difficulty in resolving conflict solely for the benefit of the client if, for instance, they are handling a number of projects that generate conflicting demands on their time and attention. Any further removal of the project manager from direct responsibility to the client makes the title difficult if not impossible to justify.

The title does not always have this reserved meaning in practice and this leads to confusion. Other titles are available which can be used to imply the orientation of the particular management activities undertaken. For instance, construction manager, contract manager, design manager are roles that are often designated project manager. The activities implied by such titles do not necessarily have the client's interest as their main concern. It must be added that they do not, of course, deny satisfying client objectives as one of their objectives. The increasing use of design-and-build structures confuses the issue somewhat as a design-and-build company will usually designate its own person as project manager but this role is distinctly different from the client's project manager (either in-house or in a consultancy role) as its focus is not solely on the client's objectives.

To complete the array of management activities in the construction industry, it should be recognised that general management of the contributing firms will also be taking place, the objective of such activities being the effectiveness of the firm.

Projects, firms and clients

Conflicting objectives

The work of firms in the construction industry and its professions presents two types of management issue: the problem of managing firms and that of managing projects. This leads to a rather complex kind of matrix management structure, shown in a much simplified form for a conventional arrangement of contributors in Fig. 1.1. This diagram is greatly simplified because it implies that the three projects are each being undertaken by the same professional practices, general contractor and subcontractors. In practice, of course, this is rarely the case. Normally there will be different mixes of professional practices, general contractors and subcontractors on each project. Even if the private practices are the same, by using competitive tendering it is very unlikely that the general con-

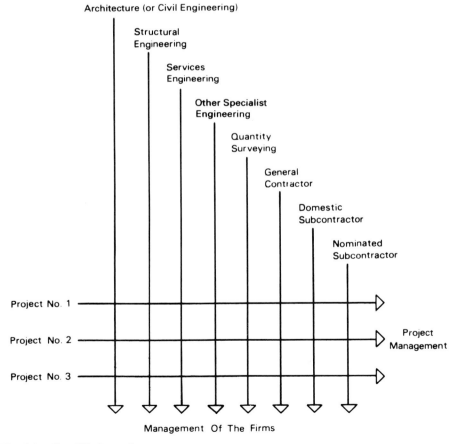

Fig. 1.1 Simplified matrix management structure.

tractor and subcontractors will remain the same. Such a lack of consistency of contributors makes it extremely difficult to improve the effectiveness of the project management process. Not only do firms have to get used to each other at both a corporate and individual level, but they are also unlikely to invest much time and money in making the process more effective when they know that any temporary management structure that they establish may only occasionally be used in a similar form again.

Construction projects are therefore undertaken by an amalgam of firms, which change from project to project. The firms involved in each project are independent companies, which are organisationally interdependent in terms of the project. This situation creates a potential for conflict between the needs of each firm and of each project. Each firm has objectives which are expressed in terms concerned with the efficiency of the firm, such as:

- increasing productivity
- improving service
- maintaining existing clients
- attracting new business.

The major purpose is to improve effectiveness and hence service and profits. Professional practices would claim to be less entrepreneurial than contracting organisations, but nevertheless conflicts between the needs of individual firms and the needs of projects will still arise. For instance, what does a firm do if there is a choice to be made because of limited resources between progressing an urgent matter for an existing client and undertaking a piece of work that could clinch a commission with a new client? Similarly, what would a contractor do if faced with a choice between keeping a piece of equipment on site to be used to keep a project on programme and removing it to another site in order to increase the profit on the second site, knowing that liquidated damages are unlikely to be claimed on the first site?

The objectives of project management, which ideally should also be the objectives of the firms involved in the project are, as has been said, the objectives of the client. These will relate directly to the project and will be:

- functional satisfaction
- aesthetic satisfaction
- completion on time
- completion within budget
- value for money.

Where, then, does the responsibility lie for ensuring that the project's objectives are met? The professional practices, particularly architects and civil engineers in a conventional arrangement, would say that it rests with them. But who, then, is to resolve the conflicts that may occur in a manner which is to the benefit of the particular project? If the practices are to do it, can they be sufficiently unbiased to resolve conflict to the benefit of the project to the extent to which the client may require?

The matrix structure using independent firms seems to need the responsibility for project management to rest in a firm or individual who is independent of the potential conflicts within the contributing firms. But may not such a firm also be faced with similar conflicts if dealing with a stream of projects.

Ideally it seems that project management should be exercised by the client organisation itself, and this reflects the need for clients to be very close to the organisation and implementation of their projects. However, many clients do not have the expertise to manage their own projects. This, therefore, is the dilemma for clients and for project management. Clients should be concerned to ensure that the design of organisation structures for their projects recognises and seeks to overcome such potential conflicts.

These issues raise the question as to whether the construction process is unique. It may not be unique in terms of its organisation problems, but it is perhaps unique in that the problems pervade all levels of construction activity and in many countries are firmly rooted in the historic development of the industry and its professions.

Interestingly, recognition of the project management process as a suitable subject for formal study and research emerged from complex projects outside the construction industry (Morris 1994), for example in connection with the defence/ aerospace programme in the USA and in other industries facing complex demands that require inputs from a range of independent firms, yet the situation they faced has been present in construction certainly since the last century.

It has taken the complexity and constraints of today's world, together with initiatives from outside the industry, to focus attention upon the way projects are managed as a possible means of finding solutions to some of the problems the industry faces. This reflects much of what Marian Bowley (1966) deduced about the inertia within the building industry and professions which has stifled innovation.

Organisation patterns

Frequent reference has been made to the conventional arrangement of independent professional firms and contractors selected in competition. However, variations in organisation patterns have grown substantially in recent years through the growth of, for example, interdisciplinary group professional practices, design-and-build companies, and joint ventures for overseas contracts all of which should overcome some of the problems associated with the variation in professional firms contributing to projects. Similar advantages could also accrue to interdisciplinary public sector groups but, on the other hand, if a public service department provides only one type of professional service for their project (e.g. architecture) and uses professional firms for other services, the problems of organisation may be even more complex.

Different organisation patterns are also generated by the way in which the general contractor and subcontractors are selected. A range of alternatives is available, for instance, design-and-build, two-stage tendering and negotiated contracts and it is necessary to ask whether a proposed contract generates the organisational form most likely to achieve the client objectives. It has been found that it is unlikely that there is one 'best' form of contract for construction projects (Nahapiet & Nahapiet 1985). One then has to add to this what is probably the most significant variable: the vast range of client types served by the industry.

Variety of clients

Clients vary in many ways. Of particular importance is the variety of objectives that clients seek to satisfy. Differences in this respect are particularly marked between private and public sector clients, and overseas and multinational clients may have objectives rarely encountered in home markets.

The variety of objectives is compounded by the range of uncertainty of clients' objectives. The construction industry and its professions have to be skilled at translating such variability in a way which enables them to produce projects that satisfy their clients. They have to deal directly with their clients and in order to do this, and obtain and interpret instructions properly, they need to understand how their clients' organisations operate as the organisation structures used by clients vary considerably to reflect the needs of clients' major activities. As everyone, either individually or corporately, is a potential client for construction work, the construction industry and its professions could be called upon to work with every or any possible organisation configuration. The industry and its professions need to understand how organisations work in order to organise themselves and also understand how their clients' organisations work, so that they may be in the most advantageous position to interpret and implement their clients' objectives.

The demands that both private and public clients place upon the construction process are frequently complex and uncertain. This simply reflects the complexity and uncertainty of the modern world, as demonstrated by contemporary economic, social and environmental issues. The construction industry and its professions are themselves also subjected directly to such forces.

When establishing temporary management structures for construction projects within such an environment, the industry has available a range of organisational approaches but has historically tended to adopt a conventional solution. Such a standard answer cannot be expected to solve problems as complex as these. The professions and industry are now more readily developing approaches to the design of organisation structures that are tailored to satisfying specific client objectives, and take advantage of the range of temporary management structures available. What is needed is a framework for designing the most appropriate structure.

The contribution of organisation structure

It is important to put the orientation of this book about organisation structure into a broader management context. There are many factors other than organisation structure that have a significant bearing upon the performance of an organisation. However, organisation structure is a particularly important aspect as, if properly designed, it allows the other aspects to function properly.

This is not to say that, if an organisation is inappropriately designed, it will not perform adequately, as people have the ability to construct informal organisation structures that circumvent the formal structure often to the benefit of performance. However, a strong informal structure can work against organisation co-ordination and control. The ideal is when the organisation is sufficiently well designed that it does not generate an informal structure. Such an outcome would mean that the organisation is designed to meet its specific objectives and that the participating members would have confirmed that, in their view, this is in fact the case.

An appropriately designed organisation structure for a project will provide the framework within which the other factors that influence the effectiveness of the project management process have the best chance of maximum performance in the interests of achieving a client's objectives. For the purpose of construction project management, the major internal factors influencing the effectiveness of the management process can be considered to be:

- behavioural responses
- techniques and technology
- decision making
- organisation structure.

These aspects are interrelated and interdependent, as illustrated in Fig. 1.2.

The project management process is also subject to external influences. These comprise all elements outside the process which, if they change, demand a response from the project management process if it is to remain effective. Examples are economic forces, which may affect the client and modify his objectives for the project, and legal forces, which may require changes to the design, e.g. revised building regulations.

Behavioural responses

The behavioural factor consists of the characteristics of the individual members of the organisation as reflected in their motivation, reaction to status and role relationships and their personal goals and values. It therefore determines the attitude they have to their work on the project and to the work of others. Attitudes are significantly affected by external influences (e.g. the views of society) in addition to being influenced by the other aspects of the management process, for example Loosemoor (1994) finds that high reciprocal interdependency of tasks

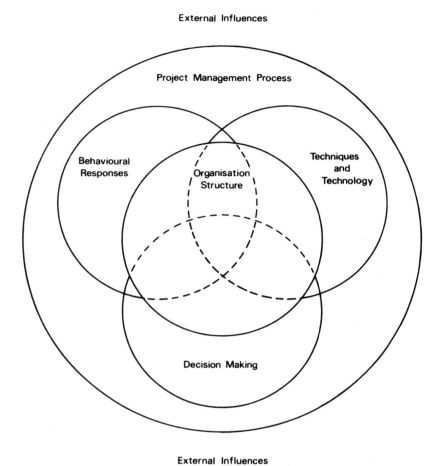

Fig. 1.2 Factors in the project management process.

frequently found on construction projects forces people to find solutions to problems. Behavioural responses have particular significance for construction project management because of the sentience of the various professions and skills involved, many of which have strong allegiances and view projects from very different positions, as illustrated in Fig. 1.3. It is a factor that can have a significant impact on the effectiveness of the project management process.

Techniques and technology

Techniques and technology are the tools used by the members of the organisation to produce the building or other construction work. The quality of the tools they use is determined by the knowledge the project team have of the techniques and

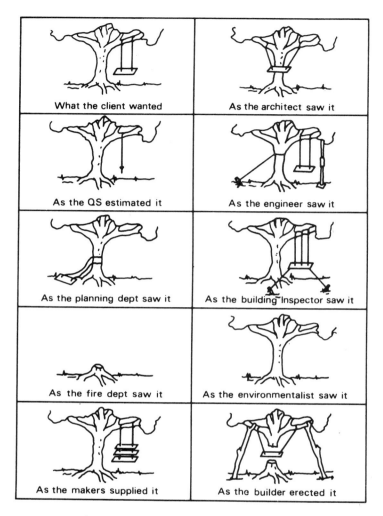

What the client wanted

As the architect saw it

As the QS estimated it

As the engineer saw it

As the planning dept saw it

As the building inspector saw it

As the fire dept saw it

As the environmentalist saw it

As the makers supplied it

As the builder erected it

(Original idea and sketches by Dave Taylor)

Fig. 1.3   Perspectives of the contributors.

technologies available and their skill in using them. The interdependency of the organisation structure and the techniques and technology used is based upon the need for the organisation to be structured in such a way that the appropriate techniques and technologies are drawn upon and used at the correct time in the process of designing and constructing. As a corollary, the techniques and technologies adopted may demand a certain organisation of the contributors to make their use effective. The techniques employed and the way in which they are put together by the project management process are fundamental to achieving clients' objectives. They encompass evaluation, appraisal and control methods, contractual techniques and approaches to design, as well as the techniques of constructing the project. Particularly important for construction projects is the project information facility employed and how this relates to the organisation structure.

Scott (1992) believes that three dimensions of technology – complexity, uncertainty and interdependency – are most important in explaining differences in structural characteristics of organisations. Technical complexity leads to structural complexity and increased professionalisation, uncertainty leads to lower formalisation and decentralisation of decision making and complexity and uncertainty to greater interdependency needing higher levels of co-ordination. From this analysis the links of technology to organisation structure and decision making are clear to see, particularly in construction.

Decision making

Although the quality of decision making is vital for all organisations, it acquires special status on construction projects. The complexity of projects is reflected by the large number of specialists who contribute to the decision-making process. It is therefore closely related to the organisation structure, which determines how people work together to produce the output that forms the basis on which decisions are made.

In moving towards the completed building, the process is characterised by a series of 'pinch points' through which it must pass if it is to make progress. At each of the points, a decision has to be made which could include the option to abort the project. The process of making these decisions will be managed by the project management process as a whole and will be based on output generated by the contributors working within the organisation structure using techniques and technologies. The client and managing executive will take many of these decisions based on the advice of the specialist contributors. The significance of decision making is that it should be interrelated with the organisation structure in such a way that advice is received by the decision maker from the appropriate contributors at the appropriate time.

Organisation structure

The organisation structure of a particular construction project is a subset of the project's management process. It structures the relationships of the members of the organisation and hence influences their responses to the demands placed on them. It establishes the way in which advice is generated for decision making and the use of techniques and technology in the process. It should be designed to allow these factors to be integrated.

The managing executive of the project should be responsible for designing the organisation structure and then should provide the integrating activities that weld the parts into a unified whole. The managing executive then provides the dynamism required to make the whole process seek to achieve the client's objectives.

Organisation theory and project organisations

Organisation theory recognises that professional organisations are distinctive. As Scott (1992) states:

> 'Certainly the most elaborate and intricate organisational arrangements yet devised for coping with high orders of complexity and uncertainty in production systems are to be found in the *professional organisation*.'

However, he continues by focusing on what he terms autonomous professional organisations and heteronomous professional organisations. The former are those in which the professionals have responsibility for their own goals and the establishment and maintenance of their performance standards. Examples are design firms, law firms, medical staff in hospitals and academics in universities. The heteronomous types are when the professional staff are subordinated to administrative staff and have relatively little autonomy. Public agencies fall into this category, such as schools and welfare agencies.

Scott recognises that autonomous professionals use a number of organisational forms, one of which is the project team. He recognises this in hospitals and university research projects. He concludes that:

> 'Organisations that in one way or another utilise lateral relationships as legitimate avenues of information and influence flows constitute the new generation of organisational forms. As we have attempted to illustrate, a number of different lateral structural arrangements are in use – including project teams, matrix structures, organic or clan systems, and professional organisations. All move us away from unitary hierarchial arrangements, "beyond bureaucracy" or "from bureaucracy to adhocracy".'

What is significant here is that the types of structure which have been used in the construction field are now seen as the 'new generation of organisational forms' by organisation theorists. This shows the need for an organisation theory related directly to the forms used for construction projects. Whilst elements of general organisation and management theory are relevant and useful to project management, that which is so needs to be carefully distilled and developed before it is able to provide insights which contribute to making project organisations more effective. The distinctions are highlighted by Thompson (1991):

> 'The temporary nature of the project team and the need to define and achieve specific project objectives against a demanding timescale, together with the high level of risk and expenditure encountered on many projects, will demand a

style of project management that is likely to be more dynamic than that of corporate management.'

Against this backcloth, it may seem surprising that project organisations have been able to design and construct projects reasonably successfully for hundreds of years and particularly so in the more unstable environment of this century. But construct them they have, nevertheless this does not mean that they cannot do so more effectively. The emergence of a theory of project management is a necessary basis for increased effectiveness in which organisation theory will play a significant part. Such theory will emerge in part from the general management literature but also more particularly from experience in practice and from original thinking specifically about project organisation.

Project organisation in context

This book is primarily concerned with the criteria against which the effectiveness of the design of organisation structures of construction projects can be judged. However, the systems approach to organisation design which is developed does not ignore the other major factors that influence the effectiveness of the project management process. Rather the systems approach provides the core to which all other elements relate. Such elements include the behavioural/psychological approaches to management particularly relating to motivation, organisation culture, strategic management, decision-making techniques and many other specific aspects of management, all of which make significant contributions to an understanding of the management process and have many publications devoted to them.

Chapter 2
Evolution of Project Organisation

Introduction

The way in which construction projects are organised in different countries has evolved from traditions and conventions laid down in each country many years ago. The traditions and conventions of the UK have had a particularly wide significance as they have been exported to many parts of the world over the last two centuries. A brief account of project organisation evolution in the UK may help to explain the position reached in trying to develop more effective ways of managing construction projects. It will have been paralleled in many other countries.

Origins

The way in which contributors to construction projects organise themselves in the UK has its origins in the Middle Ages, but the original patterns have been influenced significantly by the increasing complexity of the conditions in which building has taken place and by the construction industry's attempts to cope with the prevailing conditions.

The surviving records of building in the Middle Ages are for prestigious structures and mainly show that a master mason was responsible for acquiring and organising labour and material and for the technicalities of construction on the basis of an outline from the client. Alongside master craftsmen there often existed clients' representatives, many of whom did not have practical experience of building but who were among the few people who were literate and numerate. They were expert administrators and went under a variety of titles, such as surveyor, clerk of works and sacrist. The client would pay directly for the labour and material consumed.

This direct method predominated until towards the end of the seventeenth century, although there is evidence as early as the beginning of the sixteenth century of work being let on a contract or 'bargain' basis. This basic pattern probably had many variants as the recorded titles of people are confusing and the relative responsibilities difficult to determine.

The eminence of master masons led to the most eminent being appointed King's Mason with responsibility for oversight of the king's palaces and castles. They also acted as advisors on a number of projects, in a role akin to that of architect in later years.

The relatively stable conditions in which the 'building industry' existed in the medieval period did not create conditions for change in the organisational pattern of building work until demand for building began to rise in the sixteenth century, when the distinctive role of the architect began to emerge and more work began to be awarded on a contract or 'bargain' basis. Engineers were more concerned with mechanical devices for military purposes than with buildings, but through their work on fortifications and castles their influence on buildings began to be felt although master craftsmen developed their own empirical engineering and jealously guarded this knowledge.

The period from the sixteenth century to the Industrial Revolution saw many changes, which had profound effects on the organisation of building projects. England had become a principal trading country of the world and travel had awakened interest in the buildings of ancient Greece and Rome, leading to a demand for such designs. This led to the clearer identification of the role of the architect, and the associated complexity resulted in an increasing tendency to let building work on a contract basis, although 'architects' also often acted as developers. Further impetus to change was created by the Great Fire of London, which led to the 'measure and value' method of settling payments and the employment of separate measurers. There still appears to have been little application of formal engineering to building, although some road and bridge building took place.

The great surge in the demand placed upon the construction industry was generated by the Industrial Revolution. The accompanying prosperity created demands for housing, to accommodate both workers and owners, and for buildings for the new industries. The demand for improved transportation led to the development of new engineering and building techniques, and to further industrialisation and demand for buildings. In response to such demands, new materials were being developed, which allowed new building techniques to be devised.

These activities created a concentration upon the specialist skills of the members of the building industry. The importance of the engineer emerged; there was the further separation of the architect and builder as specialists; quantity surveying skills were more firmly identified; and engineering was subdivided into civil, mechanical and electrical skills. However, this was an incremental process and specialists often acted in dual capacities.

The new complexity of the conditions within which construction work was executed, with greater emphasis on economy, value and prestige, the complexity of new building materials and technologies and the developing skills of the building industry specialists themselves, created the need for greater specialisation among them. These pressures led to the establishment of societies for the discussion of common problems. Architectural clubs were formed in 1791, but clubs for civil engineers had been set up as early as 1771. In 1834 clubs were established for surveyors and for builders. Subsequently, to protect themselves from economic pressures on the one hand and from the unscrupulous on the other, the clubs developed, in the nineteenth century, into professional institutions as the means of defining their position and creating their public image through the acquisition of royal patronage. This further emphasised the separation of the skills associated with construction and so reinforced allegiance to specialist skills rather than the industry as a whole, and created the basis from which today's conventional organisational structure for construction projects had grown.

The period from the late nineteenth century to the First World War saw a continuing rapid increase in the growth of the building industry. This was accompanied by the rise of the general builder for both speculative and contract work, and the parallel emergence of specialist craft firms. This occurred in response to the need for organising ability and financial strengths required for the process of urbanisation and industrialisation.

The architectural profession was moulded on the social and aesthetic pattern of the eighteenth century, when architecture was considered one of the arts with the actual construction of the building being secondary. By the late nineteenth century, the idea that there should be any connection between architects and the mass of industrial buildings and working class housing seems to have been generally disregarded. Architects were, by then, concerned primarily with prestigious buildings. These attitudes were reflected in the 1887 supplementary charter of the Royal Institute of British Architects, which laid down that no member of the Institute could hold a profit-making position in the building industry and retain his membership.

This separation of architects and builders was accompanied by further separation of architects and engineers. The development of industrialisation and the position adopted by architects decreed that industrial building was the province of engineers but, at the same time, engineers were commonly employed to advise on the structure of architect-designed buildings in addition to their core work on infrastructure projects. Hence, architects were technically dependent upon engineers, but engineers were not dependent upon architects. Significantly, engineers did not exclude themselves from being principals of engineering or building firms. Further separation occurred when, in 1907, the Royal Institution of Chartered Surveyors instituted the Contractors' Rule, which prohibited its members from being employed by construction firms.

Bowley (1966) describes the pattern that emerged as 'the system' and believes that it had acquired a strong flavour of social class distinctions, architects being the elite. Engineers were associated with trade and industry, surveyors were on the next rung of the social hierarchy and builders were regarded as being 'in trade'. She believes that, as a result, aesthetic and technical innovations in the late nineteenth and early twentieth centuries were completely out of step with each other, which inhibited the development of the major technical innovations of steel-framed and reinforced concrete structures *vis à vis* other countries and created a conservatism in the construction professions.

Building activity between the First and Second World Wars was much greater than before 1914. However, there were no important changes in the way in which the design and production of buildings were organised, although the efficiency of site operations was enhanced, particularly through mechanisation. The period was one of consolidation of the main professions through the establishment of professional qualifications tested by examination and of codes of conduct, which raised their status and reinforced adherence to the established pattern of project organisation.

The lack of innovation in building in Britain in this period was brought about primarily as a result of the lack of a built-in mechanism in the organisation of design and construction that could create the necessary stimulus. The innovations that did take place tended to be outside the industry, particularly in the orga- nisation of the building materials industry and in the materials themselves. In addition, there was great concern with housing needs and the switch from commercial speculative development to public development reinforced the pre- vailing pattern of organisation as this work also used the same structures. The pattern of organisation of design and construction does not appear to have been fundamentally questioned during this period, as reflected in the list of official government publications, none of which were concerned with organisation but which concentrated mainly on materials and housing.

Present-day organisation arrangements for building projects and attitudes to innovations within the industry still reflect, to a degree, the conservatism gen- erated by patterns laid down before the Second World War. However, following a succession of official reports on these topics, the professions and industry have responded to an extent to the demands of an environment far more complex than that in which the patterns were originally established. The dramatic develop- ments in transportation, communications, health care, manufacturing technol- ogies and the associated economic, social and technological order have been important forces for client-led change in the construction industry.

A perspective of contemporary influences

The Second World War and post-war activity

The impetus to innovation provided by the Second World War was dramatic and focused upon the need for economy in labour and reduction in the use of materials in short supply. This need was demonstrated by the rapid adoption of prestressed concrete, prefabricated buildings and the tendency to replace steel with reinforced concrete. Wartime also generated the first governmental enquiry (HMSO 1944) directly concerned with the organisation of building work, which was the forerunner of reports that questioned the suitability and efficiency of the prevailing organisation of the construction process, but which focused on building rather than civil engineering. Nevertheless, this report accepted the established patterns and concerned itself, primarily, with tendering methods and arrangements for subcontractors.

Following the Second World War, the demands placed upon the building industry rapidly increased in complexity. The pressures created by the need for rebuilding in the aftermath of war were followed by an acceleration in complexity of demand through the development of the Welfare State, which required new and more advanced buildings. Also, the increased sophistication of industry required increasingly sophisticated buildings and there arose the need to redevelop cities to cope with a more technological age. One of the driving forces behind these demands was the increase in the relative importance of government-sponsored buildings and the consequently greater involvement of government in building. The tendency to build larger production units, arising from the development of large-scale organisations in industry, was another important force.

In spite of the substantial changes in demand placed upon the industry, the pattern of organisation of projects remained largely unaltered. Increased government sponsorship of building projects served to reinforce allegiance to the traditional pattern by the need for public accountability, which was seen to be satisfied by competitive tendering on finished designs. Nevertheless, there were some innovations in organisation patterns through the use of negotiated tenders and 'design-and-build', but the resistance to change of the established pattern is illustrated by the reluctance of public authorities to adopt selective, as opposed to open, tendering even though this had been strongly recommended in the Simon Report (HMSO 1944) and again in the Phillips Report (HMSO 1950). Other developments were concerned with improving the effectiveness of site operations, particularly through prefabrication and in-house construction.

However, the need for greater co-operation began to be recognised following the Phillips Report, which commented upon the ease with which variations could be introduced during construction, the problems created by drawings issued late, the extensive use of nominated subcontractors and the desirability of establishing a common basic education for all those involved in the design of buildings and their production.

Increasingly, discussion centred upon the need for greater co-operation between all parties to the construction process. This was further stimulated by the greater need for engineers to be involved in the more complicated buildings being demanded, the necessity for reliable cost control and an increase in the number of large building firms.

The difficulties of the traditional pattern of organisation in coping with the demands of modern construction, which were evident between the wars, were greatly intensified after the Second World War, but the greater spirit of co-operation within the industry that had begun to emerge took place against the backcloth of the existing traditions and was not concerned with a fundamental reappraisal of the structure that had been established. This situation was reflected in the next major official enquiry, the Emmerson Report in 1962, (HMSO 1962) which reiterated the findings of the previous two post-war reports regarding the need to improve co-ordination of the members of the building team.

The significant reports of the 1960s

Although it was concerned with supply and demand in the construction industry, standards of training, research and technical information, the Emmerson Report is particularly significant for its observations on relationships within the construction professions and industry, and with clients, and in connection with the placing and management of contracts. It identified a common criticism of the construction process as the lack of liaison between architects and the other professions and contractors, and between them and clients. It commented, 'In no other important industry is the responsibility for design so far removed from the responsibility for production.' The report pointed out that although a common course of initial study for designers and producers of buildings had been recommended in 1950, no practical steps had been taken by 1962. Emmerson came to the conclusion that there was still a general failure to adopt enlightened methods of tendering in spite of the recommendations of earlier reports. His recommendations in this respect led directly to the establishment of the Banwell Committee later in 1962, to consider these issues in more detail.

Of the official reports, the Banwell Report (HMSO 1964), and its review Action on the Banwell Report (HMSO 1967) had a significant impact upon the construction industry and its professions. A particular concern of the report was the unnecessarily restricted and inefficient practices of the professions, leading to over-compartmentalisation and the failure of the industry and its professions to think and act together. The 1967 review found some progress on preplanning projects but noted that the professions had done little to 'de-restrict' their practices. The review was encouraged by the increase in selective tendering and urged further consideration of serial and negotiated tendering. The Banwell Report also related to civil engineering as well as building.

The Emmerson and Banwell Reports brought into sharp focus the need to reform the approach to the organisation of construction projects. They were accompanied by other reports making similar points. At the time, construction project management was seen to be a passive procedural activity but the movement towards a more dynamic integrated approach was being suggested by Higgins and Jessop (1965) in a pilot study sponsored by the National Joint Consultative Committee of Architects, Quantity Surveyors and Builders. They clearly identified that the problems of communication in the building industry were created to a large extent by attitudes and perceptions about the values of contributors to the building process. They were probably the first to suggest that overall co-ordination of design and construction should be exercised by a single person (or group). Concurrently, a review of the construction industry by the

National Economic Development Council (1964) was calling for improvement in the management of the construction process, and the co-ordination of activities of the members of the construction team and the administrative framework within which they were working. A rather rhetorical report by the Institute of Economic Affairs (Knox & Hennesey 1966) was also condemning the restrictive practices of the professions.

This spate of activity and concern with the performance and organisation of the industry and its professions marked the beginning of a self-examination. It was induced, to a large degree, by external pressures that reflected the greater complexity of the influences at work upon the industry and its clients. The economic expansion of the early 1960s and rapidly developing technology and changing social attitudes were manifested in demands for more complex and sophisticated projects and a more economic utilisation of resources. These forces were transmitted to the industry through its clients and also directly affected its techniques and attitudes, but such self-examination was likely to be slow when undertaken in the presence of the polarisation of skills and attitudes inherent in the professional structure that had emerged over the preceding century.

The reorientation of management studies of the construction process that had begun to take place is well illustrated by the Building Research Station's report in 1968 (Bishop 1968). This identified the fact that up to that date most of the work of the BRS had been concerned with the management of building sites and building firms but noted that future work would be concerned more with the management of the total building process.

The project manager and other organisation initiatives

During the 1960s and subsequently, progress was made in developing collaborative work and skills, and in instituting procedures that provided a variety of organisational patterns, particularly in connection with the introduction of the contractor at various stages of the design process. However, there was still a need in official reports in the 1970s (National Economic Development Office, 1975, 1976, 1978) to stress that more attention should be paid to structuring and managing project organisations to create conditions for co-operation between contributors. Each of these reports recognised the distinctive nature of the project management process and the role of the project manager, and reflected the changes in attitudes and views expressed since the mid-1960s. They arose from the distillation of the professions' and industry's experience of working with novel forms of organisation. The 1976 report recognised the need for further study which would analyse existing patterns in the use of alternative methods of organisation of the design and construction process.

The external pressures that caused the professions to reconsider organisational arrangements for projects were accompanied by challenges from the Monopolies and Mergers Commission in relation to their codes of conduct and fee scales. Further pressure was brought to bear through the definition and development of project management concepts and applications in other industries and the recognition by project management theorists that the concepts and techniques are applicable to construction. The professions' and industry's response to these influences reflected the manner in which the traditional structures emerged. Each sector pursued its own approach to project management while recognising rather reluctantly that the role of project manager was not the right of any one profession.

A reflection of the unco-ordinated empirical evolution of project management as an activity separated from design skills is given by the number of definitions that emerged. The Chartered Institute of Building's (1979) paper identified thirteen definitions. It commented that the confusion of terminology and usage was unsatisfactory, and proposed a further definition! It was, perhaps, to be expected that those writing on such an important emerging idea, which was contrary to their traditional backgrounds, should seek to express their ideas in their own words. However, this resulted in a range of definitions that tend to reflect the particular background and experience of the writer rather than a generalised definition of the concept.

The empirical nature of publications on project management was reflected in their emphasis on defining the jobs to be done by a project manager at various stages of a particular project rather than identifying the concept and process of project management. Nevertheless, such publications have been useful in emphasising the patterns that can be adopted with advantage to the client.

Against this background of pressure for change in organisational approaches were a number of project-based initiatives. The project manager idea was only one such idea which was used to cover a range of organisational patterns. Others included *management contracting, design and construct contracts* and *construction management* all of which seek to increase integration (particularly of the contractor) and which may or may not incorporate a project manager but which do not necessarily overcome the polarisation of professional attitudes.

The 1980s saw some recognition by the professions and industry of the forces working upon them. There was a shift from the government-sponsored reports of the 1960s and 1970s to initiatives from the private sector reflecting the shift in the political climate as a more pragmatic position was adopted. Clients were increasingly making their voices felt. Carpenter (1981) was typical of clients, stressing that the industry frequently adopted inappropriate organisation structures and the British Property Federation (1983) came up with its own system to impose on an industry which was not changing itself sufficiently quickly. Government reflected this pragmatism with so called 'client guides' to procurement (Department of Industry 1982; National Economic Development Office 1985) and practical comparisons of different approaches to development with the emphasis on speed of construction (National Economic Development Office 1983, 1987).

Alongside these developments, the original edition of this book in 1984 and other texts on construction management (Bennett 1985) were stressing the need for organisation structure to be tailored to suit the particular project. In parallel, developments on project information co-ordination resulted in the publication of the Co-ordinated Project Information (CPI, 1987) documents which aimed to improve the co-ordination of drawings, specifications and bills of quantities. The success of the Co-ordinating Committee which consisted of **RIBA, RICS, BEC** and **ACE** is indicative of the change in attitude that was taking place within the professions.

Within this climate the UK professions and industry began to implement more flexible approaches to the management of projects. As a result, the 1980s saw more adaption to suit the needs of clients and a more ready acceptance of alternative arrangements than in any other decade. Nevertheless, Mohsini and Davidson (1992) estimate that 80 per cent of all building projects in the USA were still procured by traditional processes and the figure will probably be similar in the UK although Bresnen and Haslam's (1991) work implies that it may be somewhat less at about 70 per cent.

Following the spate of reports during the 1970s and 1980s the process drew breath until the Latham Report (Latham 1994) which reinforced the pragmatic tone of the 1980s rather than the more philosophical tenor of earlier reports. Whilst its focus was predominantly on contractual matters and their impact on conflict, payments and cash flow, it nevertheless found room for an important section on project management. The changes which have taken place since the 1970s are clearly recognised by the report's comment that 'there is increasing (if sometimes reluctant) acceptance that Project Management, and a separate discipline of Project Manager, are permanent and growing features of the construction scene'. The report continues by recognising that the manner by which project management can be provided takes a great many forms which may or may not require someone with the title of project manager. Recommendations are made requiring a clearer definition of the role and duties of project managers.

Nevertheless, comments on project management are rather overshadowed by the contractual matters and also by the high profile target of 30 per cent real cost reduction of construction costs by the year 2000. However, a focus of the working group set up to implement the cost reduction initiative is on organisational matters to answer such questions as: How do you decide what to build? What is the best way to set up a team? Whilst much progress has been made in adopting flexibility in structuring and managing projects it appears that some fundamental questions still require attention.

The Latham Report begins its executive summary by stating that previous reports on the construction industry have either been implemented incompletely, or the problems have persisted, and many of Latham's recommendations reflect this as they are not new, although suggestions for achieving them may be. A recommendation which is new and important to effective project management is that the use of Co-ordinated Project Information should be a contractual requirement. By contrast the report again recognises the need for common professional education in just the way that the Emmerson Report of 1962 did which was itself reflecting Phillips' Report of 1950. the progress, nearly 50 years on, has been minimal.

Techniques for project control have been available for many years, for example, program evaluation and review technique (PERT) and critical path methods (CPM) both came about in the 1950s, but their rate of application has been variable depending upon the inclination of the team leader who traditionally was the architect. Indeed project management is sometimes seen as a collection of techniques rather than the framework in which they are applied. Peter Morris (1994) writing in the mid-1990s had cause to comment:

> '(Project Management) is widely misperceived as a collection of planning and control techniques rather than as a rich and complex management process. Indeed many of the project management specialists themselves perhaps do not fully recognise the real scope of the discipline.'

Although useful for learning from the experience, such reviews, reports and developments do not address the need for a generalised conceptual framework that allows identification of the features of significance in the construction process as a basis for designing organisation structures that take account of them. A conceptual framework is needed that allows project management functions to be identified so as to reflect the demands of different projects and resolve the differences that appear to exist when identifying functions from an empirical base.

Relevance of systems theory

In recent years there has at last been some valuable initiatives in response to the pleas of the successive official reports for greater co-ordination, but there remains a resistance to change which reflects strongly entrenched attitudes and loyalties. Against such a background, progress has been incremental. However, an equally significant inhibition to progress has been the lack of a fundamental framework of organisation theory relating to construction projects against which experience of the various organisational initiatives can be measured and compared.

Systems theory provides the opportunity to develop such a framework. *General systems theory* (GST) originated in the biological sciences, but its originator (von Bertalanffy 1969) has acknowledged its general applicability, which he considers encompasses business organisations. It has been usefully applied to organisational problems in industries other than the construction industry.

The attraction of systems theory as a medium for identifying a conceptual framework for the management of the construction process lies in the basic premise that a system is an organised or complex whole: an assemblage or combination of things or parts forming a complex or unitary whole, which is greater than the simple sum of the parts. The systems approach stresses the contribution of the interrelationships of the parts of the system and the system's adaption to its environment in achieving its objective. Early recognition of the value of the application of systems concepts to the organisation design of the construction process was demonstrated by Morris, Napier and Handler.

Peter Morris (1972) developed an approach to studying integration of the participants at the design–construction interface of construction projects. Morris's work supported the systems approach in that he found that organisation theory, especially when employed in the context of a systems framework, could be used to describe and explain the nature of the management process of construction projects.

A further application of the systems concept was made in Sweden by Napier (1970). In this work he attempted to gain an understanding of the problems of the Swedish building industry as a whole as a basis for the design of systems for the future. Handler (1970) was principally concerned with the building as a system. This concept was developed by reference to GST, by drawing an analogy between a living organism and a building.

Later work of significance has been undertaken by, for example, Ireland (1985), Rowlinson (1988) and Hughes (1989). All were underpinned by a systems perspective applied to different issues of importance to the construction industry. Ireland used the systems model of organisations as a general paradigm and from that base investigated the relationship between the use of particular managerial actions and the achievement of project objectives. Rowlinson attempted to indicate some rational basis for the choice of procurement form for the management of construction projects by identifying those factors which significantly affect project performance. This was done with particular reference to the distinction between design-and-build and traditional procurement forms. Hughes adapted the model described in this book to analyse the management of public sector projects to identify the elements of the project management process which contributed or detracted from their success.

These studies illustrate the potential for the application of systems theory to the building process. Each study took a different perspective, but employed the same basic concepts. The fundamental premise of systems theory stresses interrelationships and is as concerned with the links between the parts of the system as with the parts themselves. The problem of how to make the links work effectively is essentially the problem of project management. In order to apply these ideas to the construction process to the greatest benefit, it is necessary to take as broad a perspective of the process as possible from conception of the project to completion and even beyond.

Chapter 3
Organisation and the Construction Process

Introduction

In a business setting managers depend on effective organisations for achieving
their objectives. But there are also organisations in many contexts other than
business, for example public, social and sports organisations. In all cases the
members seek their objectives through their organisations. Organisations are
ubiquitous. It is hardly surprising therefore that studies of organisations have
arisen from many traditions, from both practice and academia, and from such
disciplines as sociology, psychology and management and, more recently, the
feminist movement.

For project managers the problem is to distil from what has been described by
Koontz (1961) as the management theory jungle, those aspects of the work on
organisation which are relevant and useful to managing the construction project
process more effectively, particularly as the jungle has become much more dense
since Koontz's description over 30 years ago.

The objective of this chapter is to briefly explore the range of perspectives on
organisation and identify those attributes which are valuable in a project management context, before developing them further in future chapters. Classifying
organisational approaches is fraught with difficulty as each phase does not have a
distinct beginning and end. Rather, organisational approaches evolve over time
as each new approach draws something from previous approaches, so the distinctions apparent in the presentation here should not be seen as absolute divisions.

The classical approach

The classical approach stems from the first wave of writing on management in the
early part of this century and is characterised by the work of Taylor (1911) and
Fayol (1949 trans.) and was the foundation of management practice up to about
1950, although many will say that it still continues to be so in many cases today.

Taylor laid the foundation of 'scientific management'. The approach determined that it was possible to scientifically analyse and structure the tasks to be

performed so that the maximum output could be obtained with the minimum input. This approach meant that people were perceived as machines and efficiency was the sole criteria of success. The outcome of such an approach led to increasing specialisation of the workforce. Managers' activities were also seen to be governed by set processes and procedure as much as the workers.

Fayol, whose work did not become freely available until after translation in 1949, was nevertheless influential before then as others in the USA had developed his ideas. They developed 'principles of management' which were concerned with such things as pyramidal structure, unity of command, line and staff, the scalar chain and span of control. The primary element was the pyramidal organisation structure and the idea that authority is delegated downwards. Division of labour was advocated so that the sub-goals of the various units would add up to overall organisational goals and co-ordination would be handled through the management hierarchy. The principles emphasised formalisation and specialisation and were in this way complementary to and supportive of Taylor's scientific approach. It may be that in those days the construction professions recognised, through their experience, that such approaches were not really appropriate to the management problems of construction projects and, for this reason, did not fully develop management in their training at an early date.

What emerged from this classical view of management and hence organisations was a deterministic perception. The 'principles' were held to be universal truths about how management should be undertaken and the only way to manage business activities or processes. Hence an extremely rigid view of how to organise emerged. As with many originators of theory, Fayol did not intend this outcome. It was the users who followed his 'principles' slavishly (and many still do so today). Fayol made the point that they were not rigid. Maybe his mistake was calling them 'principles':

> 'The soundness and good working order of the body corporate depends on a certain number of conditions termed indiscriminately principles, laws, rules. For preference I shall adopt the term principles whilst disassociating it from any suggestion of rigidity, for there is nothing rigid or absolute in management affairs, it is all a question of proportion. Seldom, do we have to apply the same principle twice in identical conditions; allowance must be made for different changing circumstances.' (Fayol, 1949 trans.)

His contemporaries made the point even more tellingly:

> 'Students of administration have long sought a single principle of effective departmentalization just as alchemists sought the philosopher's stone. But they have sought in vain. There is apparently no one most effective system of departmentalism.' (Gulick & Urwick 1937:31)

It has taken many years for these early signals to be taken on board by both practitioners and academics but developments in organisational theory have eventually reflected these views.

Of great influence at this early stage in the development of thinking on organisations was the work of the German sociologist Max Weber (1968 trans.) He worked independently of Taylor and Fayol and adopted a different stance. Rather than focusing on how to improve organisations he took a far more academic approach of seeking to describe the characteristics of newly emerging bureaucratic structures. The characteristics he identified were generally compatible with other ideas of the time such as specialisation, hierarchy, etc., but were developed in much more depth. His primary focus was on organisations as power structures in which control is achieved through an organisation hierarchy. Discipline was the keyword which required the exact execution of orders from above.

The classical approach to organisations and management was therefore seen as essentially rigid and originated from military and church models which strongly influenced the way in which the early managers organised. It did not make explicit the effects of the human component and external influences on organisations.

The behavioural approach

Serious study of people in organisations did not begin until it was explicitly recognised that informal organisations existed in parallel with formal organisations. Recognition of informal organisation structures alongside the formal, and the shortcomings of classical organisational theory, saw the emergence of the behavioural schools, which believed that the study of management should be centred on interpersonal relations or that it should be seen as a social network.

Informal structures exist alongside formal organisational structures because people cannot be treated as machines. Their behavioural responses to their position within a formal organisation cannot be expected to subscribe to the predetermined manner in which they are expected to perform. Hence an informal structure will arise. How different this structure is from the 'official' structure will depend on many factors not least of which will be how well the formal organisation has been constructed. Contributing significantly to this scenario is goal complexity within organisations. There is often a disparity between the official goals of an organisation and the goals actually pursued which then govern the behaviour of the participants. In such situations the existence of unofficial goals creates an informal structure intended to achieve them. Hence the behaviour of the members of an organisation cannot be constrained within an inappropriate organisation structure. The recognition of this phenomenon challenged the classical approach to organisations. A wide range of researchers contributed to the development of this new approach, far too many to cover here, but a few important ones are briefly discussed to give a sense of their thinking.

The famous Hawthorn experiments conducted by the Harvard Business School in the 1920s and 1930s laid the groundwork for much which followed. As Scott (1992) states:

'The experiments served to call into question the simple motivational assumptions on which the prevailing rational models rested. Individual workers do not behave as "rational" economic actors but as complex beings with multiple motives and values; they are driven as much by feelings and sentiments as by facts and interests; and they do not behave as individual, isolated actors but as members of social groups exhibiting commitments and loyalties stronger than their individualistic self-interests.'

As a result, many sociologists and social psychologists devoted their attention to examining how people in organisations actually behaved, how they actually related to their supervisor, subordinates and peers and what were the factors which motivated members of organisations. McGregor's (1960) work encapsulated much of this work and contrasted the social system approach with the classical approach through his now famous 'Theory X' and 'Theory Y' assumptions about how people behave in organisations. Assumptions underlying Theory X were:

- 'Individuals dislike work and will seek to avoid it.'
- Therefore, 'most people must be coerced, controlled, directed, threatened with punishment to get them to put forth adequate effort toward the achievement of individual objectives'.
- 'The average human being prefers to be directed, wishes to avoid responsibility, has relatively little ambition, wants security above all.'

By contrast, under Theory Y:

- Most individuals do not 'inherently dislike work ... the expenditure of physical and mental effort in work is as natural as play or rest'.
- 'External control and threat of punishment are not the only means for bringing about effort toward organisational objectives.'
- The most significant rewards are those associated with 'the satisfaction of ego and self-actualization needs'.

Barnard's (1938) significant earlier work also falls within this general category. He stressed that organisations are co-operative ventures which integrate the contributions of their members. He also dealt significantly with authority in organisations (referred to in Chapter 9) which is reflected in his view that goals are imposed from the top down whilst their attainment depends on the bottom up. Much of the work on leadership (see Chapter 10) also stemmed from the behavioural approach.

The behavioural movement began in the 1930s and continued into the 1970s with perhaps its zenith in the 1950s. However, questions were raised as to whether the field of human behaviour is equivalent to the field of management as the nature of an organisation's goals and consequential necessary tasks of organisations can make this approach impracticable. Also research on the basis of empirical evidence has failed to show relationships between behavioural aspects of organisational members and productivity (Scott 1992). Eilon (1979) refers to the 'myth of self-actualisation'.

Between and within the classical and behavioural schools are a wide range of approaches. The main criticism now levelled at them is that at the time each was offered as the one best way to organise. Subsequent organisational structure thinking denies such an assumption but believes that each may have something to offer within a systems framework.

The socio-technical approach

The Tavistock Institute of Human Relations, London, undertook a series of studies in the 1950s and 1960s which contributed significantly to the development of systems theory and its application to business organisations and the construction industry.

They developed what at the time was a distinctive research approach in that they proposed that the unique feature of business organisations is that they are both social and technical systems. The socio-technical approach emphasises that the needs of both the technical and social aspects should be served by organisations. This view contributed to combining many of the previous approaches some of which considered only technical needs whilst others considered only social needs. Bennis (1959) labelled the former 'organisations without people' and the latter 'people without organisation'. As Scott (1992) notes, the goal should be one of 'joint optimisation' of the needs of both the technical and social systems, since the two systems follow 'laws' and their relationship represents a 'coupling of dissimilars' (Emery 1959). From this view emerged the recognition of the impact of environments on organisations. Scott (1992) also notes that the Tavistock work is essentially European with little similar work having been carried out in the USA.

The Tavistock group undertook an important study of communications in the construction industry (Higgins & Jessop 1965; Tavistock Institute 1966) which identified the main features of the technical system as 'interdependency and uncertainty'. In terms of the social system they highlight the mismatch between the traditional organisational arrangement with the architect as designer/manager and the organisational separation of production undertaken by the construction company. Their report was the first to identify the need for someone in a separate project management role. Winch (1989) criticises their work on the grounds that 'nowhere in either of the two reports is there any analysis of the implications of the contracting relationship . . .' and identifies two weaknesses: 'It was developed for handling relationships within simple organisations' and '. . . the perspective is psychologically orientated, and so has difficulty in grasping a context in which the actors have differing economic interests'. Winch puts these arguments within the context of supporting a transaction cost approach to the construction firm and the construction project.

What Winch fails to do in these arguments is to separate the firm from the project. The Tavistock group was concerned with process and the needs of the fundamental organisational system for the production of a project which was satisfactory to the client. The manner by which the organisational units which carry out that process are formed is then a separate matter and the transaction cost approach may well be an elegant explanation of this structure. Nevertheless the process of creating and constructing a project has basic organisational needs which should be satisfied by the provision of an ideal organisational arrangement which a client should seek in order for their project to have the best chance of satisfying their needs.

This argument is put here as it is a recurring issue in the application of organisational theory to construction projects. The needs of the management of the process and the needs of the management of the firms should in the first instance be considered separately but with the needs of the process taking precedence so that the selection of organisational units can be taken in an informed manner. That is not to say that Tavistock should not have taken transactions costs into account, although much of the work on the link of the transaction cost approach to organisations took place much later than Tavistock's pioneering work.

The systems approach

The classical and behavioural approaches did not seem to offer much to the construction professions and industry although construction firms may have been attracted to the formality of the classical approach. As a result, traditionally there was little management included in courses for the construction professions but rather more in courses directed at the construction firms. In particular the classical and social systems schools did not appear to have anything to offer the management of the process of producing a project. The managers of the process were alienated by each school claiming to offer the only way to manage.

The management abilities of the professions and industry were, therefore, mainly acquired through the experience of managing in the real world. Although such experience is vital, its value is reduced if it cannot be gained within a conceptual framework of management theory. The ad hoc acquisition of theory has meant that generally most older professionals' and industrialists' knowledge relates to traditional management concepts. It was only with the advent of the systems approach to organisation structure that members of the construction professions and industry had a theory to which they could relate their experience of their particular industry. This means that only recently have qualified members been formally exposed to systems thinking.

The systems approach is essentially a way of thinking about complex processes so that the interrelationships of the parts and their influence upon the effectiveness of the total process can be better understood, analysed and improved. Its origins lie in the biological sciences through its founder Ludwig von Bertalanffy who devised general systems theory (GST) from his consideration of the fundamental interdependency of many aspects of science which were studied independently. He generalised his theory (von Bertalanffy 1969) to show that it was applicable and valuable to a broad spectrum of disciplines and it was subsequently applied to business organisations.

The appeal of the systems approach to the study of construction project organisations arises from its focus on how the parts of a process are dependent upon each other, as illustrated by the following definition of a system:

> 'An entity, conceptual or physical, which consists of interdependent parts. Each of a system's elements is connected to every other element, directly or indirectly, and no sub-set of elements is unrelated to any other sub-set. (Ackoff 1969).

It is clearly the case that the success of the construction process depends to a large extent upon the way in which the architect, engineer, quantity surveyor, contractors and others work together. It depends upon them perceiving the same objectives for the project and recognising that what each of them achieves depends upon what the others do. With this view they should be able to stand above the particular interests of their own contribution and see the problem posed by the project as a whole. The advent of the project manager has, to a large degree, come about as a result of the inability of the contributors to consistently achieve this, and in response to the consequent need for someone to concentrate solely upon integrating the various contributors in the interests of the client.

To understand how the building process operates as a system it is necessary to understand the distinction between closed and open systems. A *closed system* is one that does not respond to events and occurrences outside the system. It cannot adapt to changes and is therefore predictable. Machines can be considered to be closed systems in that the parts are selected to perform specific functions in a given set of conditions to produce a predetermined output. If there are changes in the conditions for which the machine was designed, the machine will not adapt to them. For example, a washing machine will not work if overloaded, a motor car will not work properly on dirty petrol.

On the other hand, an *open system* adapts to events and occurrences outside the system. These events and occurrences take place in what is known as the system's environment. This has been defined as a set of elements and their relative properties, which elements are not a part of the system but a change in any of which can produce a change in the state of the system (Ackoff 1971). An open system has a permeable boundary and there is import and export between an open system and its environment. It is therefore influencing and being influenced by its environment. An open system is dynamic and adapts to its environment by changing its structure and processes. Although stable, it is always changing and evolving and presents differences over time and in changing circumstances. A living organism is an open system and business organisations are analysed as open systems.

However, it is not as clear cut as the closed–open dichotomy implies: there is a range of other classifications. For example, a central heating system and the human body, both of which adapt themselves to changes in the temperature of their environment by internal adjustment so that they remain static, are referred to as homeostatic systems. Also, Child (1977) has described a system that exists in a protected environment in which it defends itself from having to adapt fully to its environment. Therefore the system is not fully open.

Business organisations could never have existed as closed systems. Similarly, the construction process has always been an open system. Potential clients exist in the environment of the construction process system and the system must adapt to them. It imports ideas, energy, materials, information, etc., from its environment and transforms them into its output, which is the finished construction. This is then exported to the environment, which is itself influenced by the use to which the completed project is put and by the fact that the construction is an addition to the nation's fixed capital. The process is illustrated in Fig. 3.1.

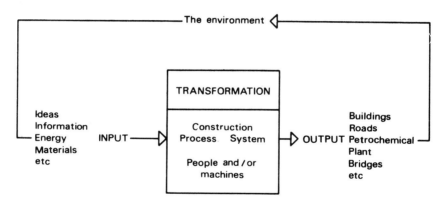

Fig. 3.1   The construction process as an input–output model.

Recognising the construction process as an open system means that the functions upon which the project management process should focus can be summarised as:

- identifying, communicating and adapting the system's objectives
- ensuring that the parts of the system are working effectively
- ensuring that appropriate connections are established between the parts
- activating the system so that the connections that have been established work effectively
- relating the total system to its environment and adapting the system as required in response to changes in its environment.

In practical terms the project manager will be concerned particularly with anticipating the chain reactions of decisions and developments that occur on the project. For example, as a result of an upturn in business, the client may decide at a late stage in the design of a project to be submitted for competitive tender that substantially more floor area is needed in the factory. This decision has to be appraised in terms of its effect on the project cost, completion time and functional efficiency, and evaluated against alternatives such as providing the additional area in a different form, e.g. leased accommodation, or by a different method, e.g. a negotiated contract for the additional area. This will require interaction between all the contributors to the project, the complexity of which will depend upon the actual stage of development of the project. The final decision will need to be taken solely in terms of the client's objectives in relation to the revised requirements. For example, completion time not only means additional construction time but also the additional time required by the consultants and the effect this may have on construction completion, particularly if this may mean 'hidden delays' to construction completion because of drawings being issued late or incomplete. The relationship between cost and completion time would also need to be appraised. All of these factors would require interaction between the contributors so that the priority of the competing demands would be resolved in the client's interest. The project manager needs to be able to anticipate the interconnectedness generated by such decisions and to manage the system with respect to them.

General systems theory developed alongside the various schools of management thought and it had an attraction for management thinking as it presented an opportunity to converge these strands of thought within an acceptable and theoretically sound framework with less rigidity and more recognition of interdependency in organisations than previous approaches allowed. The systems approach reflects the scale of interdependency created by the nature of activities to be undertaken (e.g. the design and construction of a building) and the effects upon the activities of environmental influences. It therefore discounts rigid approaches that propose one method for all circumstances. This is not to say that the systems approach discounts as irrelevant the ideas of classical management and the behavioural schools, but rather that it provides a framework for understanding and analysing organisations through their internal and external relationships, which places into context the earlier views of organisations. For example, the behaviour of individuals within an organisation remains important but it is more easily understood and relevant if it is seen within the context of the relationships demanded by the activities being undertaken and the environment within which they take place.

Contingency theory

Impetus to the application of the systems approach to management came through Lawrence and Lorsch's (1967) major study which led to the contingency theory of organisations design, which states that there is no one best way to organise but rather that organisation is a function of the nature of the task to be carried out and its environment. It encompassed many applications of systems ideas to organisations. Lawrence and Lorsch found that different environments, which generate different levels of uncertainty, require varying degrees of separation (differentiation) of organisational units (e.g. architect, engineer, contractor) and hence they require different degrees of integration.

The extent of differentiation within an organisation depends upon the uncertainty and diversity of the environment and the effect this has on the way the task has to be organised and managed. Lawrence and Lorsch state that they found that the amount of differentiation in the effective organisation was consistent with the environmental demand for the interdependence of the parts of the organisations. In developing their contingency theory they state that this starting model is complicated as soon as we move to a complex, multi-unit organisation, in which each unit strives to cope with different parts of the environment. For example, a construction project that is carried out in conditions of uncertainty and is technologically complex requires a wide range of specialist skills, which are closely dependent upon each other, in achieving a successful outcome. As soon as this happens, it introduces the complication of integrating the work of different units. Lawrence and Lorsch see the existence of an integrating unit and conflict-resolution practices as contributing to the quality of integration and in turn to overall performance. This unit has come to be represented on construction projects by project managers.

A number of other significant research studies building on systems theory led up to the contingency theory. For example, by Burns and Stalker (1966) which analysed firms in the electronics industry and identified two patterns of organisations and management. The one they termed 'mechanistic' was similar to the classical model referred to earlier. The other, termed 'organic', had a participative character. The 'mechanistic' and 'organic' structures lie at the extremes of a spectrum which illustrates the range of approaches possible. Burns and Stalker did not suggest that either was superior to the other. They concluded that, when taken in context with the task and environment being considered, one pattern will be more appropriate for the specific tasks and environment in question.

The contingency theory is a succinct summary of a great deal of detailed work that went before it. It is perhaps a reflection of the management discipline's apparent need to sum up a complex situation in just a few words. Child (1977) is critical of the contingency theory on these grounds and believes that it has not in the main recognised the organisation design difficulties which may result from the presence of multiple contingencies. He is concerned at the situation in which a configuration of different contingencies is found which are conflicting in terms of organisation design. For instance, a construction project may demand a relatively bureaucratic organisation structure to ensure accountability, but at the same time require a more loosely structured organisation to more readily allow innovation to take place. Child also questions the cost effectiveness of the additional integrating mechanisms required, as he is not convinced that there is evidence that they improve performance.

Nevertheless, even allowing for such criticisms, the systems approach as summarised in the contingency theory provides a framework for thinking about the design of construction project organisations and for analysing them, so that the effect of organisation structure on the outcome of projects can be better understood.

Strategic contingency

The strategic contingency approach adopts an open system approach as does contingency theory but it arrives at a different rationale for the structure of organisations. Contingency theory believes that managers have to respond to the environment of their organisations in designing organisations hence they are responsive to, and their actions determined by, the environment. Strategic contingency theorists believe that managers have choices and although the environment may constrain their choices to some extent it does not determine them. They recognise the role of power in determining the strategy to be adopted.

Following from this view is that, rather than being the function of task and environment, organisation structures are determined by political contests within organisations (Pfeffer 1978) leading to a framework for the power-driven political explanation of organisational structure.

Resource dependency

The resource dependency model also arises from the open system framework and can be seen to be associated with the strategic contingency approach as its primary concern is the impact of external forces on how firms organise (Pfeffer & Salanick 1978). There are two major elements. One is that organisations are constrained by and depend on other organisations that control resources which are critical to their operations and the other is that organisations attempt to manage their dependencies on external groups to acquire more autonomy and freedom. The resource dependency model sees managers making strategic choices within constraints to reduce their dependencies, which illustrates the model's similarity to the strategic contingency approach. However, the model's view is that managers do not have unbridled strategic choice as Child (1972) originally proposed in developing the strategic contingency approach but that they do exercise some discretion over how to structure organisational relationships to manage the uncertainties created by dependency which requires adjustment of inter- and intra-organisational linkages as summarised by Greening and Gray (1994).

Scott (1992) believes that the resource dependency model means that organisational participants, particularly managers, scan the relevant environment searching for opportunities and threats, attempting to strike favourable bargains and to avoid costly entanglements. All organisations are dependent on supplies and consumers but which specific exchange partners are selected and what are the terms of exchange are partly determined by the organisation itself. Astute managers acquire the necessary resources but do so without creating crippling dependencies.

Institution theory

Institution theory is concerned with the impact of institutions (government, professional groups), public opinion and pressure groups on the structure of organisations. Meyer and Rowan (1977) argue:

> 'Many of the positions, policies, programs, and procedures of modern organisations are enforced by public opinion, by the views of important constituents, by knowledge legitimated through the educational system, by social prestige, by the laws, and by the definitions of negligence and prudence used by the courts. Such elements of formal structure are manifestations of powerful institutional rules which function as highly rationalized myths that are binding on particular organisations.'

Institution theory is part of the group of ideas which stem from the treatment of organisations as open systems. It adopts a perspective which is different from the strategic contingency and resource dependency approaches in that its focus is that many of the more powerful forces in the environment are social and cultural pressures to conform to institution preferences and conventional beliefs rather than pressures to adapt to a more productive organisation. Response to such pressures is seen to provide organisations with increased legitimacy, resources and survival capabilities.

Postmodernism

Recent years have seen a collection of ideas and critiques which have challenged established assumptions about the way in which organisations are analysed. These ideas reject the idea that there is a reality which can be objectively observed and measured so that universal laws or features or relationships within and between organisations can be established.

Postmodernist conception substitutes interpretation for explanation, as described by Scott (1992):

> 'An interpretive approach to analysing social systems signifies a number of important changes. First, as Agger (1991) observes, "postmodernism rejects the view that science can be spoken in a singular, universal voice". Rather, "every knowledge is contextualized by its historical and cultural nature". Different truths are associated with differing social or temporal locations. "Social science becomes an accounting of social experience from these multiple perspectives of discourse/practice, rather than a larger cumulative enterprise

committed to the inference of general principles of social structure and organisations." A related difference: all knowledge is self-referential or reflexive; that is, we are interpreting a subject's understanding of events in the world not only in their relation to one another, but to ourselves.'

The implications of these ideas for project management in construction, if any, is not likely to emerge for many years but Green (1994) tentatively addressed the nature of social reality and its relevance to interpreting the briefing process for projects and suggested that the dominant paradigm of UK building procurement during the 1980s was based on positivism whereas the dominant paradigm during the 1990s is based on social constructicism.

Mintzberg's classification

Mintzberg's classification of the structure of organisations has a strong appeal to people in the construction industry as demonstrated by Bennett's (1991) adaptation of Mintzberg's (1989) idealised types to suit construction, to which further reference will be made later.

Mintzberg's work is based on an open systems approach incorporating contingency theory as he believes that effective organisations achieve an appropriate balance between task, environment and organisation structure but he sees his configuration approach taking it further. This he characterises as 'getting it all together', in which the elements are selected to achieve consistency. He argues that academic research on organisations has favoured analysis over synthesis and has focused on how variables arrange themselves along linear scales rather than how attributes configure into types.

His basic premise is that a limited number of configurations can help to explain much of what can be observed in organisations. His analysis of organisations prior to synthesis is extensive and insightful but there is no space to do it justice here and the reader is referred to the original work. His seven configurations do, he believes, encompass and integrate much of what is known about organisations. He emphasises that each configuration is idealised. As he says: 'a simplification, really a caricature of reality. No real organisation is ever exactly like any one of them. But some do come remarkably close, while others seem to reflect combinations of them, sometimes in transition from one to another'.

It should be pointed out that five of his configurations appeared in his book *The Structure of Organisations* (Mintzberg 1979) – Entrepreneurial, Machine, Diversified, Professional and Innovative. The other two – Missionary and Political – were added later (1989) and are distinctly different from the earlier five in that they overlay on the other five and are seen as forces in organisations rather than the form of an organisation. However, in their extreme forms they can become so strong that the organisation's structure is built around them.

The seven configurations follow using Mintzberg's own words. In the case of the original five he also includes structure, context, strategy and issues in their specification for which the original text should be consulted:

The Entrepreneurial Organisation

Nature:
  '• Simple organisations that are run by their leaders.'

Conditions:
  '• Fostered by an external context that is both simple and dynamic. It must be relatively simple (say retailing food as opposed to designing jet aircraft) in order for one person at the top to retain so much influence, and it is the dynamic context that requires the flexible structure, which in turn enables this organisation to outmaneuver the bureaucracies.'

The Machine Organisation

Nature:
  '• Supposedly the big bad guys of the organisation world, the homes of red tape and the sources of curious tales. Yet if we think of McDonald's or the Swiss railroad, a different impression develops, of organisations – when they get it right – that can be enormously efficient and can provide an unmatchable reliability of service.'

Conditions:
  '• Work of a machine bureaucratic nature is found, above all, in environments that are simple and stable. The work associated with complex environments cannot be rationalised into simple tasks, and that associated with dynamic environments cannot be predicted, made repetitive, and so standardised.'

In addition, the machine configuration is typically found in mature organisations, large enough to have the volume of operating work needed for repetition and standardisation, and old enough to have been able to settle on the standards they wish to use. These are the organisations that have seen it all before and have established standard procedures to deal with it. Likewise, machine organisations tend to be identified with technical systems that regulate the operating work, so that it can easily be programmed.

The Diversified Organisation

Nature:
  '• The waves of mergers that have taken place over the last century have led to the formation of giant corporations and to the so-called divisionalised forms of structure. The 'conglomerate' is, of course, the ultimate example of this, where a corporation doesn't much care about any relationship among its different businesses other than financial.'

Conditions:

'● While the diversified configuration may arise from the federation of different organisations, which come together under a common headquarters umbrella, more often it appears to be the structural response to a machine organisation that has diversified its range of product or service offerings. In either case, it is the diversity of markets above all that drives an organisation to use this configuration.'

The Professional Organisation

Nature:

'● It is the one place in the world where you can act as if you were self-employed yet regularly receive a paycheck. These seemingly upside-down organisations, where the workers sometimes appear to manage the bosses, are fascinating in the way they work. We need professional organisations to carry out highly stable tasks in society, such as replacing someone's heart or auditing a company's books. But as a society we have yet to learn how to control their excesses: professionals who mistreat their clients, professional organisations that mistreat their supporters.'

Conditions:

'● The professional form of organisation appears wherever the operating work of an organisation is dominated by skilled workers who use procedures that are difficult to learn yet are well defined. This means a situation that is both complex and stable – complex enough to require procedures that can be learned only through extensive training yet stable enough so that their use can become standardised.'

The Innovative Organisation

Nature:

'● Some people refer to this type as 'high technology', and to its basic orientation as 'intrapreneurship', an indication that whereas the entrepreneurial configuration innovates from a central individual at the top, this one depends on a variety of people for its strategic initiatives. These initiatives tend to be many, because what adhocracy provides is sophisticated innovation. That comes at the price of a good deal of disruption, if not chaos, and wasted resources. This type achieves its effectiveness by being inefficient. Perhaps that explains why it confuses many people: the innovative organisation may be necessary, but it is not conventional, at least not by the standards of the traditional literature of management.'

Conditions:

'● This configuration is found in environments that are both dynamic and complex. A dynamic environment, being unpredictable, calls for organic structure; a complex one calls for decentralised structure. This configuration is the only type that provides both. Thus we tend to find the innovative organisation wherever these conditions prevail, ranging from guerrilla warfare to space agencies. There appears to be no other way to fight a war in the jungle or to put the first man on the moon.'

The Missionary Organisation (Ideology)
- '• Rich system of values and beliefs that distinguishes an organisation.'
- '• Rooted in sense of mission associated with charismatic leadership, developed through traditions and sagas and then reinforced through identifications.'
- '• Can be overlaid on conventional configuration, most commonly entrepreneurial, followed by innovative, professional, and then machine.'
- '• Sometimes so strong that evokes own configurations.'
- '• Clear, focused, inspiring, distinctive mission.'
- '• Co-ordination through the standardisation of norms ('pulling together'), reinforced by selection, socialisation, and indoctrination of members.'
- '• Small units ('enclaves'), loosely organised and highly decentralised but with powerful normative controls.'
- '• Reformer, converter, and cloister forms.'
- '• Threats of isolation on one side, assimilation on the other.'

The Political Organisation
- '• Means of power technically illegitimate, often in self-interest, resulting in conflict that pulls individuals or units apart.'
- '• Expresses itself in political games, some coexistent with, some antagonistic to, some that substitute for legitimate systems of power.'
- '• Usually overlaid on conventional organisation, but sometimes strong enough to create own configuration.'
- '• Conventional notions of concentrated co-ordination and influence absent, replaced by the play of informal power.'
- '• Dimensions of conflict – moderate/intense, confined/pervasive, as well as enduring/brief – combine into four forms: confrontation, shaky alliance, politicised organisation, complete political arena.'
- '• Can trace development of forms through life cycle of impetus, development, resolution of the conflict.'
- '• Politics and political organisations serve a series of functional roles in organisations, especially to help bring about necessary change blocked by legitimate systems of influence.'

Mintzberg then took his ideas on configurations a stage further by conceiving that, rather than being a perfect fit, many organisations fit *more or less* into one of his configurations. Organisations which would be expected to be a perfect fit for one of the organisational types demonstrated anomalies, some quite limited but others quite extensive. As a result he conceived his configurations as forces which exist to a lesser or greater degree in all organisations at one time or another if not all the time. At the same time he maintained that the seven configurations demonstrated the most fundamental forms that organisations can take and which some do for some of the time. This is summarised in Mintzberg's Pentagon of Forces and Forms shown in Fig. 3.2.

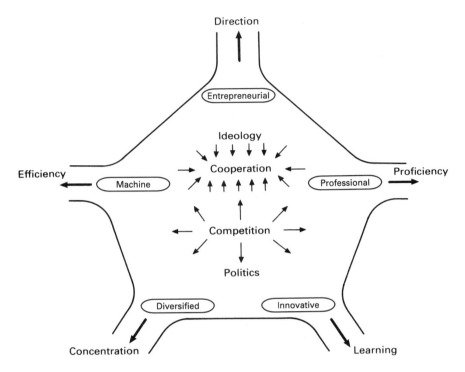

Fig. 3.2 An integrating pentagon of forces and forms. (*The Structuring of Organizations* by H. Mintzberg, © 1979. Adapted by permission of Prentice-Hall, Inc., Upper Saddle River, N.J.)

Mintzberg believes that the forces and forms as conceived by the pentagon 'appear to constitute a powerful diagnostic framework by which to understand what goes on in organisations and to prescribe effective changes in them'.

Bennett (1991) creatively uses Mintzberg's ideas and applies them to construction to provide valuable insights. He identifies three systems of organisation which he describes as idealised and unlikely to occur in their idealised forms in practice. Bennett's three configurations (which he terms gestalts) are programmed organisations, professional organisations and problem-solving organisations. They have ideas in common with Mintzberg's machine, professional and innovative configurations respectively.

Bennett defines his gestalts as follows:

'Programmed organisations are so called because their work is highly rationalised. As a consequence, the organisations set up to deal with individual projects are very simple. However, they form part of much larger organisations which take full responsibility for the design, manufacture and construction of standard buildings, bridges or other standard constructions.

The second type of gestalt is professional organisations. They are so called because their work is based on using professional skills and knowledge within the boundaries of established technical rationality. Professional construction project organisations depend on the existence of some form of traditional construction. That is a form of construction where designers know the performance they will achieve from using any particular combination of design details. Also, the local construction industry knows the nature of the work required to manufacture and construct any particular combination of design details. Local contractors know the sequence of specialist contractors required, the effective construction methods, and the plant and equipment needed, and can predict the resultant costs and times with confidence.

The third type of gestalt is problem-solving organisations. These are organisations which produce innovative constructions efficiently. They are called problem-solving because they are set up to find answers to customers' needs which cannot be met by established answers.'

Bennett makes reference to Mintzberg's ideas of organisations being subject to the seven forces as illustrated in his pentagon but believes that these ideas are not relevant to construction projects as project organisations are 'small enough and short term enough to make it possible and appropriate for them to adopt the pattern of a pure gestalt'. It is debatable whether Mintzberg's approach to the synthesis of organisational forms is transferable to construction projects. Mintzberg's work is directed at firms as organisations and the structures are essentially intraorganisational. On the other hand, Bennett is synthesising project organisations which are usually, although not always, a coalition of a large number of independent firms. The issues which manifest in project organisations are about process and different from those within firms. The firms which comprise a project coalition, of which there can be many, include the client, the professional consultancies, the contractor and subcontractors. Each of these companies can be categorised as one of Mintzberg's seven configurations or as a hybrid from his pentagon of seven forces. The nature of the project and hence the process will determine which configuration is likely to make a firm the most appropriate for each major task.

The starting point will be the client organisation which could be any of Mintzberg's configurations. Then, for example the design firm is likely to be a 'professional organisation' but may well tend towards an 'innovative organisation' and in fact should have such a configuration for demanding one-off projects requiring unique design solutions. The other professional consultancies could have similar profiles but with engineers and quantity surveyors more likely to be more biased towards a 'professional organisation'. At the construction stage of a conventional project the contractor and subcontractors will tend towards a 'machine organisation' but the nature of construction is that elements of both 'professional' and 'innovative organisations' will probably be necessary. Some of the smaller subcontractors may be mainly 'entrepreneurial organisations'. Depending on the contract strategy, the construction firms may need to be more strongly biased towards 'professional' and 'innovative organisations', for instance in design-and-build and build-operate-transfer projects.

The issue is therefore the nature of the project organisation which overlays and comprises a range of firms each of which has a structural orientation which suits its particular contribution to the project. It should be added that the selection of the firms to be part of the project coalition should ideally be made so that their configuration reflects the needs of the particular project. Straightforward projects would be suitable for professional consultancies with a 'professional organisation' configuration and the contracting firms with a 'mechanic organisation' configuration.

The issue of the distinction between managing firms and managing inter-firm project organisations is one which is not addressed in the mainstream of management literature and will be returned to at the end of this chapter.

The transaction cost approach and organisational economics

The transaction cost approach emerged from the seminal work of the economist Coase (1937) on the theory of the firm in which he advanced his theory of the existence of firms. He generalised this by stating (Coase 1988) that:

> 'The existence of transaction costs will lead those who wish to trade to engage in practices which bring about a reduction of transaction costs whenever the loss suffered in other ways from the adoption of those practices is less than the transaction costs saved.'

Thus, put in simple, but inelegant terms, those who wish to trade will do so through firms which exist in the market or through creating a new firm to carry on the trade or through expanding their existing firm to cope with the additional trade, depending on which alternative generates the lowest transaction costs. Coase therefore went on to say that, in the absence of transaction costs, there is no economic basis for the existence of the firm and that the limit to the size of the firm is set where its costs of organising a transaction become equal to the cost of carrying it out through the market. Markets and firms are alternative instruments or governance structures for completing a set of contracts.

Coase originally used the terms 'the cost of carrying out a transaction by means of an exchange on the open market' and 'the costs of market transactions' which was ultimately termed 'transaction costs' by others. Dahlman (1979) described them as 'search and information costs, bargaining and decision costs, policing and enforcement costs'. Put very simply in management terms, Coase's approach says that a firm will go to the market for the goods, services, etc., it needs rather than provide them from within the firm when the transaction costs in doing so are lower, and vice versa. It is not an argument that the market should always be the option, which is the way in which his work has often been interpreted by others. What he does say is 'What my argument does suggest is the need to introduce positive transaction costs explicitly into economic analysis so that we can study the world that exists' but he believes this has not been the effect of his work and makes a plea for objective evaluations based on his work.

Williamson's (1975, 1979, 1985) work stemmed directly from Coase and gave a particular perspective to the transaction cost approach. It has been Williamson's work which was picked up in the management literature rather than Coase's. Williamson (1979) argues that the difference in governance costs stems from the motivation, cognitive limitations and the moral character of the people involved. His assumptions about the people involved include material self-interest, bounded rationality and negative opportunistism. As a result transactions cannot be achieved unless both parties have confidence in the arrangements which overcome the negative effects of these assumptions. If they are not overcome, transaction costs will rise and may make other organisational forms more attractive. Griesinger (1990) believes that 'if researchers can identify those dimensions of organisational transactions that most affect the costs of governance, it should be possible to choose the most economical organisational form under any circumstances'. Williamson (1979) believes that frequency, uncertainty and asset specificity were the three critical dimensions. Asset specificity means the degree to which the value of an investment is dependent on maintaining the specific relationship in which it is involved or used.

Griesinger (1990) identifies that:

'The transaction cost approach has been applied at three levels: (a) the boundaries of the firm, that is, determining which activities should be governed internally and which should be contracted outside; (b) the overall structure of an enterprise and the relationship of its operating parts, that is, distinguishing between various corporate forms such as the functional, holding company, and multidivisional designs; and (c) the internal organisation of human assets, that is, matching the internal governance processes to the attributes of the workers, their groups, and their tasks. For each case the logic is the same: first, the defining transactions are identified and classified according to the underlying dimensions of frequency, uncertainty, and asset specificity, and next, efficient governance relationships are sought that will protect against opportunism.'

Griesinger and others have traced the exchange perspective to Barnard (1938), Simon (1947) and March and Simon (1958) and to the resource dependency perspective and state that it has become increasingly pervasive throughout social science but he, like many others, does not refer to the fundamental work of Coase. What has emerged is a body of theoretical knowledge labelled by many as organisational economics (OE) (Hesterly *et al.* 1990), in the formulation of which

economists have increasingly focused on issues traditionally the remit of organisation academics.

Agency theory is also seen to be an element of organisational economics and is
well stated by Donaldson (1990a):

> 'Agency theory holds that many social relationships can be usefully understood
> as involving two parties: a principal and an agent. The agent performs certain
> actions on behalf of the principal, who necessarily must delegate some
> authority to the agent (Jensen & Meckling 1976). Since the interests of the
> principal and agent are inclined to diverge, the delegation of authority from the
> principal to the agent allows a degree of underfulfillment of the wishes of the
> principal by the agent, which is termed agency loss. Agency theory specifies the
> mechanisms that will be used to try to minimise agency loss in order to
> maintain an efficient principal–agent relationship.'

The conjunction of economics and organisational theory has, not surprisingly,
caused conflicting views to emerge on the substance of organisational economics.
Whilst the transaction costs approach has enriched understanding of organisations and has for the first time established why organisations exist and how they
are formed (Hesterly *et al.* 1990), nevertheless the assumptions underlying its
more recent applications to organisations (as opposed to Coase's original trea-
tise) has created much debate not better recounted than in a discussion in 1990
(Donaldson 1990a,b; Barney 1990). Donaldson argued from the perspective of
management theory and Barney responded from a organisational economics
perspective. Donaldson objects to the assumption that managers behave
opportunistically (which is defined as the inclination to lie, cheat, steal, shirk,
etc.). Barney argues that this is not a required assumption of organisational
economics. He argues that the cost of distinguishing between opportunistic and
non-opportunistic behaviour is in fact a transaction cost.

Donaldson asserts that organisational economics adopts a narrow basis of
analysis whereas management theory adopts a holistic system level analysis.
Barney agrees on the reductionist stance of organisational economics but argues
that many management themes are similarly reductionist. Organisational economics' view that managers always act in their own self interest is refuted by
Donaldson. Barney agrees and believes that organisational economics requires
the ideas of more sophisticated models of motivation.

Barney does not believe that these differences will prevent an integration
between the models but that much of the division is the fear of some traditional
management scholars of economics imperialism and that this fear is not
unfounded. In a response to Barney, Donaldson (1990b) believes that 'organisational economics should be encouraged as one class of organisational theory
but without being allowed to dominate other theoretical approaches and with an
eventual aim of integration in a unified organisation theory'.

It seems that the debate, whilst interesting, is also to an extent artificial. If the
argument returns to Coase's theory of the firm it follows (Hennart 1994):

'In conclusion, a firm will earn rents if it can reduce its organising costs over those which are shouldered by its competitors. A firm can use three strategies. The first one consists in reducing the cost of organising interactions within the firm by devising and enforcing better employment contracts. A second strategy is to increase the efficiency of external contracts. If a firm can organise co-operation with other firms at lower costs than its rivals, then it can share in the additional gains of trade and garner additional rents. Thirdly, a firm may be more skilled than its competitors at assessing the relative costs of each. It may therefore earn rents by shifting from an external transaction to an internal one, and vice versa, if by so doing it lowers its organising costs.'

Management theory should seek to identify how the various units (internal or external) which generate transaction costs can organise and manage themselves in order to reduce transaction costs and so identify the organisational form which provides the best strategy. In so doing, firms will be managed in a particular manner which encompasses, to a lesser or greater effectiveness, organisation, leadership, motivation and all the other elements which contribute to management. What the transaction cost approach contributes is how (as a product of the employment of management theory) the competitiveness of the various organisational forms compare. Organisational solutions will be preferred (and hence, in the long run observed) only if they offer efficiency gains over other alternative arrangements (Hesterly *et al.* 1990). Management theory contributes to the efficiency of organisation and is therefore surely capable of being integrated with organisational economics.

For all these seeming divisions Smith *et al.* (1995) see that issues of co-operation are fundamental to management success and of increasing importance. Informal and formal co-operation occur. The former is concerned with behavioural norms and is traced from the early work of Fayol (1949 trans.) through to Lawrence and Lorsch (1967) and their focus on integrating mechanisms. They also see co-operation as formal when characterised by contractual obligations and formal structures of control with a formal hierarchy and rules and regulations, which may provide a perspective for integrating organisational economics and management theory. They also point out that co-operation between organisations has only recently been seen as important. This view will be surprising to people in construction where it has always been essential to the success of projects.

A reconsideration of the human side of organisational economics within the framework of transaction cost analysis, particularly relating to the economic role of co-operation raised by Barnard (1938), is introduced by Griesinger (1990). He believes that co-operation is an interpersonal resource with economic advantages for many organisations and what he terms 'betterment' for most participants, again pointing towards integration of economics and management theories.

There have been few applications of the transaction cost approach to the construction industry. Gunnarson and Levitt (1982) and Reve and Levitt (1984) analysed construction contracts as ways of governing construction transactions but also expanded their analysis into other client–consultants–contractor relationships. The analysis was not quantitative but more in the nature of a general discussion on the issues. Winch (1989) criticised their work on the grounds that they take the object of analysis as the project rather than the firm. He believes that the project is not an economic entity and does not make resource allocation decisions but that the firms do. He considers that the key question is why construction firms choose to contract for construction services rather than provide these services themselves. However, the analysis is able to focus on the project as a whole if the object of the analysis is taken as the client and the question is why the client contracted for the provision of the project rather than providing it itself. Other analyses with clear objects can then stem from this fundamental analysis.

Winch (1989) then himself applies the transaction cost approach from Williamson's perspective (1975, 1981a, 1981b) to construction by asking the question why construction firms choose to contract for construction services, rather than employ the capacity to provide these services themselves. In his analysis he argues that hierarchy (that is, retaining work in-house rather than subcontracting) would significantly reduce management overheads in the industry, citing project uncertainty, project complexity and post-contract bilateral monopoly as the elements leading to the hierarchy response. He states that in particular, the designer/main contractor and main contractor/specialist subcontractor transaction interfaces could be beneficially governed by hierarchy rather than the market. He goes on to argue that:

> 'So far as the client was concerned, responsibility for the effective completion of the contract would rest with one firm on a "turnkey" basis. Games of tag between the architect, engineers, main contractor and specialist contractors would cease as they were resolved within the integrated construction firm by, in Williamson's terms, "fiat". Hierarchy would also economise on the directly quantifiable element of market transaction costs – the costs of preparing bills of quantities and other contract documents; multiplicated estimating effort by subcontractors; external arbitration in disputes; and multiplication of contract management effort by designers, quantity surveyors, and contractors as each party tries to deflect opportunistic behaviour by the others. Thus, hierarchy would significantly reduce management overheads in an industry which has seen a considerable rise in such overheads during the 1970s.'

He then asks why there has been little shift towards hierarchy in the British construction industry. On the contrary he believes that the trend has been towards greater market governance and he cites management contracting and the subcontracting system as illustrations.

On the other hand, Chau and Walker (1994) found that the transaction cost approach was a powerful tool in explaining the extent of subcontracting in Hong Kong's construction industry. Their work illustrates the basic point that the choice of contracts is not random. Rather, it is predicated on the attempt to minimise transaction costs. The number of contracts, the availability of market information about material costs, time and expenditure in negotiating and drafting contracts, quality assurance, and contract enforcement are the institutional costs that arise in subcontracting. That the parties involved are prepared to pay these costs voluntarily indicate that they are more than offset by the savings in other types of institutional costs under an alternative arrangement.

Two alternative arrangements which are extreme cases of direction by the market and by hierarchy were examined. On the one extreme, the end-users of the product (construction work) contract directly with individual workers, specialists, and machinery owners who contribute to the construction of different parts of the structure. In this case, there will be no subcontractors, contractor and even developer. All end-users simply contract with each other to share part of a development project and they in turn contract directly with individual resource owners (together with a professional co-ordinator if necessary) to construct the building. The cost of such an arrangement is however exceedingly high due to the large number of contracts involved. The end-users do not normally know the technical details of the structure and therefore to reach the price of a component of the structure is difficult. There is also a major difficulty in separating the contribution of different resource owners to the production of an identifiable component of the structure. It is also costly for the end-users to discover their preference.

On the other extreme, the main contractor (or developer in the most extreme case) can contract directly with the workers and instruct/direct the workers what to do with no or little reference to the market prices of their outputs, as in the case of the classical Coasian firm. Since the workers are not paid according to their output as in the case of the subcontracting but paid by some proxy of their output, such as hours worked, the cost of monitoring their performance becomes very high. Another important cost is that of planning the construction activities to minimise the worker's idle time. The fact that neither of these two extreme arrangements are adopted in reality suggests that the type of transaction costs mentioned in each case are higher than the transaction cost of adopting a subcontracting arrangement.

The most important costs of subcontracting, i.e. the costs of specifying the subcontracting packages and agreeing on their prices, are relatively lower than the major transaction costs identified in the above two extreme cases. This is confirmed by the fact that all subcontracted work is relatively discrete and easily identifiable. Since both contractor and subcontractor are specialists who process market information about the output of the subcontractor, it is relatively easy for them to agree on the price.

These analyses make a valuable preliminary contribution to the implications of the transaction costs framework for explaining the structure of the construction process. However different perspectives of analysis are necessary. Analysis from the client's perspective may produce transaction costs which argue for a different structure than one produced using the construction firm as the focus and conclusions may pull in opposite directions. However, some convergence may occur as illustrated by the increase in design-and-build type contracts in recent years.

The major problem of applying the transaction cost approach is the difficulty of measuring/ranking actual transaction costs. Only if empirical work is undertaken can the true potential of the transaction cost framework be realised and arguments such as those above accepted or refuted. However, the operationalisation of the transaction cost approach involves costs which may be extremely difficult to measure accurately particularly if such costs encompass intangible costs such as those associated with motivation and, for example, those incurred in distinguishing between opportunistic and non-opportunistic behaviour. Translating sound theory into useful applications will not be easy due to problems akin to those experienced with cost-benefit analysis.

The expectation that the measurements needed to establish whether organisations should exist as markets or hierarchies can be made is perhaps unrealistic. What determines organisational form is likely to be perceptions of costs in the minds of those who make decisions on the type of organisation to use. Hence many organisational forms can exist, particularly in construction, which are efficient to a lesser or greater extent and which respond to the transaction cost framework to a greater or lesser extent. What the limits are we do not yet know.

Projects, firms and process

Whilst much of contemporary organisational theory stems from systems theory – contingency theory, resource dependency theory, institutional theory and the work of Mintzberg – its focus has been on the processes taking place within firms. It seeks to explain how firms should organise and how they should be managed in order to be effective. It was not until the development of the transaction cost framework and organisational economics that the most fundamental question – why do organisations exist? – was answered. Even more recent has been the recognition by mainstream organisation researchers of interorganisation (firm) co-operation.

Whilst organisation theorists may argue the merits of different approaches, many of the approaches, taken together, provide a basis for analysing construction project organisations. What appears to happen in organisation theory is that the distinction is not drawn between interfirm processes and firms but this is fundamental to the understanding of construction project organisations. Even Winch (1989) when considering the firms which make up project organisations sees Williamson's interpretation of the transaction cost approach as an alternative to the socio-technical analysis of the Tavistock school and to differentiation and integration concepts rather than complementary to them.

The development, design and construction of a construction project is a process which requires a range of diverse skills which is well reflected by Mintzberg (1979):

> 'Every organised human activity – from the making of pots to the placing of a man on the moon – gives rise to two fundamental and opposing requirements: the division of labor into various tasks to be performed, and the co-ordination of these tasks to accomplish the activity. The structure of an organisation can be defined simply as the sum total of the ways in which it divides its labor into distinct tasks and then achieves co-ordination among them.'

A project organisation comprises tasks and co-ordination. On the basis of organisation theory, and simply stated, each of the units which carry out the tasks will arise as an organisational entity each resulting from the application of the transaction cost approach. Each of these tasks may be contained in a separate firm or combined within a firm. Each of these firms will require to be effectively organised for the task it is required to carry out. The nature of the organisation of each firm will be determined by the contingency theory or a derivative. Each of the firms will subscribe to one of, or a combination of, Mintzberg's configurations or result from the effects of his forces. For example, the design firms may be professional/innovative organisations and the construction companies entrepreneurial/machine organisations as they cope with different tasks and environments.

Mainstream organisation theory deals with the above scenario reasonably well but in construction those 'effectively managed firms' will have to co-operate in an interorganisation structure – the project team. This will itself require a structure which is appropriate to the task of producing the project as a whole rather than each part of the project.

Williamson's assumptions about the human component greatly shape transaction cost theory. Those of material self-interest and negative opportunism sit uncomfortably against the need for high level co-operation and integration of the firms in the project process. Transaction cost theory as derived by Coase is a better basis for explaining the existence of the firms whilst the manner by which they co-operate and integrate is better explained by the contingency theory of organisation and the behavioural and motivational elements of management theory. Without these characteristics the achievement, not only of the world's spectacular structures, but also the many commonplace yet complex buildings and infrastructure projects, would never have been successfully completed. Lansley (1994) remarks that when presented with descriptions of the three types of organisation – market, hierarchy and clan – those with extensive experience of construction often remark that the parties of the typical building project come together through the processes of the market, are expected to operate according to the rule of the hierarchy but, in order to achieve a successful project, have to adopt the characteristics of a clan!

As this book is concerned with the process of management of the project as a whole and not the management of the individual firms except insofar as this affects the management of the process, the rest of the book stems from the application of systems theory, particularly contingency theory, to the construction project management process.

Chapter 4
Systems Thinking and Construction Project Organisation

Introduction

For many years project management has been synonymous with a hard system approach. That is, it has:

'... emphasised quantitative techniques in project planning, scheduling and control. Project-network analysis using PERT or CPM, earned-value measurement, variance analysis, cost-estimating techniques, risk analysis, Monte Carlo simulation, sensitivity analysis, cost modelling, and, lately, expert systems are almost synonymous with "modern" project-management approaches and techniques.' (Yeo 1993)

The focus on techniques, particularly critical path analysis, is reflected in much of the literature and major books on the subject (Turner 1994). It is remarkable that project managers have taken so long to recognise that, in spite of the many advanced techniques, projects in many fields have still been subject to large cost overruns and delays. The engineering background of many project managers has meant that they instinctively adopt a numerical approach to solving problems even when problems do not respond to such advances. Even in the 1990s project managers were behaving as though soft systems approaches had only recently been discovered although they had been around for many years:

'The soft systems approach is, above all, concerned with human behaviour in organizations, and requires radically different skills in its application: a basic intellect, an ability to see more than one point of view, to think logically, to advocate and to communicate become more important than applying scientific methods, searching for some elusive truth and reducing all problems to rigorous mathematics.' (Daniel 1990)

Morris (1994) goes so far as to rename 'project management' as 'the management of projects' to emphasise that the management of design, technology, political forces, cost-benefit, finance and more which are contained in the latter are substantially more important than techniques.

The hard system approach originated in the operational research field which has had limited application in construction. Project management in construction has therefore to an extent picked up a soft systems approach somewhat more readily than other fields of application of project management and has applied systems ideas to the design of project management organisation structures. Even so the majority of books on project management in construction are also technique orientated and there still remains a hard system ethos due to the engineering background of many in construction.

The greatest benefit of the application of systems thinking to construction project management is the structuring of organisations in such a way to achieve the client's objective. Only if this is done in the first instance can hard systems approaches to planning and decision making, and even soft systems approaches to structuring problems, be successful. The use of such techniques and approaches within an inappropriate organisational structure can only mean that their results will be inappropriately applied as the systems objectives and/or structure will be unsound.

Systems theory and its derivatives as applied to the management of business organisations, together with organisational economics, are important to an understanding of how construction projects are organised and managed. Organisational economics help to explain why the firms which undertake the work needed to produce the project are formed with their particular configurations of activities. Systems theory provides a framework for understanding how the process of undertaking the tasks needed to produce the completed project within its environment should best be carried out. That is, it helps to explain how the project management process should be structured and hence how the various firms and other organisational units are integrated into a unified process for the production of the project. Systems theory is also relevant to how the firms and their organisational units organise internally and hence how their response to their task and their environment impacts on the project management process as a whole.

The application of systems theory to business organisations has focused on the contingency theory of organisation which states that organisation is a function of the task to be carried out and the environment within which it has to be performed. Misinterpretations of this approach, as with the transaction costs approach, are not uncommon, for example Rouleau and Sequin (1995) state:

'In this approach, the representation of the individual is simplified, since both employee and manager are relegated to the background, to be replaced by the profit advantages inherent in environmental constraints, which are principally of an economic nature, and which determine the structural arrangement of the organization.'

Whilst the task and environment determine the structural arrangements, profit advantage is not the sole purpose of organisational arrangements. The purpose of the structural arrangements is organisational effectiveness which may be measured by many criteria other than profit. In business organisations market share, long-term relationships and status can all qualify the profit motive. The main point is that objectives should be clearly defined so that organisational structure can be designed to meet them.

Systems concepts

It is worthwhile to examine the relevance of the major systems concepts to the construction process to see if they can be used to give a better understanding of construction project management. Embedded in the systems approach are a number of common characteristics of systems which, although couched in systems terminology, can be interpreted in terms of the construction process. The universality of the systems approach is demonstrated by the way in which people from diverse industries have found the concepts acceptable and useful when they have worked them through in their own terms.

The basic distinction between open and closed systems has been dealt with earlier. The response of the open system to its environment being its major distinguishing feature. The original rational models of organisations (e.g. Taylor, Fayol, Weber) and the natural models of, for example, Barnard, Mayo, as defined by Scott (1992), were closed system views as they did not formally incorporate interaction with the environment within their systems.

Traditional management 'theory' had a fixed view of management. It evolved around 'principles', which were held to be universal truths about how sound management should be undertaken. The principles were considered to be the only way to manage business activities or processes, irrespective of the external conditions in which they were carried out. Many of the earlier concepts in the social sciences and in organisation theory were therefore closed system views because they considered the system under study as self-contained. They concentrated only upon the internal operation of the organisation and adopted highly structured approaches. Nevertheless, elements of the rational and natural models continue to be useful in explaining organisation structure and behaviour if incorporated within an open system framework. As a result the open system framework is seen as the most powerful paradigm for integrating other theories of organisation.

Whilst the basic dichotomy of closed and open systems is sufficient for a basic understanding of business organisations it is important to recognise that real-life businesses are to a greater or lesser extent open or closed depending upon the way in which they react to their environment. It is probably only a theoretical possibility that a business system could be entirely closed (it would die) or entirely open. Understanding of systems is aided by Boulding's classification of systems by their level of complexity as summarized by Scott (1992):

'(1) *Frameworks*:   systems comprising static structures, such as the arrangements of atoms in a crystal or the anatomy of an animal.
(2) *Clockworks*:   simple dynamic with predetermined motions, such as the clock and the solar system.

(3) *Cybernetic systems*:   systems capable of self-regulation in terms of some externally prescribed target or criterion, such as a thermostat.
(4) *Open systems*:   systems capable of self-maintenance based on a throughput of resources from its environment, such as a living cell.
(5) *Blueprinted-growth systems*:   systems that reproduce not by duplication but by the production of seeds or eggs containing preprogrammed instructions for development, such as the acorn – oak system or the egg – chicken system.
(6) *Internal-image systems*:   systems capable of a detailed awareness of the environment in which information is received and organised into an image or knowledge structure of the environment as a whole, a level at which animals function.
(7) *Symbol-processing systems*:   systems that possess self-consciousness and so are capable of using language. Humans function at this level.
(8) *Social systems*:   multi-cephalous systems comprising actors functioning at level 7 who share a common social order and culture. Social organisations operate at this level.
(9) *Transcendental systems*:   systems composed of the "absolutes and the inescapable unknowables".' (Boulding 1956)

As Scott points out the nine levels are not mutually exclusive as each higher level system incorporates the features of those below it. It is possible to analyse level 8 using any of the levels lower than 8, 7 using any levels lower than 7 and so on. Boulding believes that 'much valuable information and insights can be obtained by applying lower level systems to higher level subject matter'.

A further basic concept is that of organisations as hierarchies of systems. Hierarchies in this sense does not refer to levels of authority in the classical management meaning but to the arrangement of sub-systems, systems and supersystems. Each system is part of a larger system and also comprises other systems. If the physical entity of a building is seen as a system, it can then be conceived as part of the collection of buildings on the road on which it is situated, which is the building's senior system, the buildings on the road are part of the district system, which is a sub-system of the town system which is a sub-system of the country and so on. To understand a system it is necessary to look outside the system at the system in which it is contained which can also be conceptualised as part of its environment depending on where the system boundary is drawn.

A further basic idea is that of loosely coupled systems. The fundamental idea of a system is that it consists of interdependent parts. The impression given is that they are tightly and strongly linked and change in response to each other. This can be seen as an overgeneralisation. It is possible that the parts are relatively weakly connected and capable of autonomous action not requiring a response for other parts to which they are connected. This view is analogous to the fully closed, fully open perceptions of systems' reactions to their environments. The parts of the system may be on a scale from tightly to very loosely coupled but nevertheless they are interdependent. The parts of the project management system tend to be tightly coupled in terms of technical dependency but even this can vary. The degree to which they are organisationally coupled in terms of organisational development, long range objectives, etc., is certainly variable depending to a large extent on the organisational configuration adopted, as required by the task and the environment of the project.

Objective

A system has an objective. The objective should be stated as clearly as possible and developed as further information becomes available. The manager of an organisation must ensure that all members of the organisation are aiming for the same objective, and must attempt to resolve conflicts where they occur. Many business organisations find it difficult to identify their objectives explicitly but it is an important task of the manager to identify as clearly as possible the objectives of the organisation, communicate it to the members and gain their acceptance. If the objective is unacceptable to members, it will be difficult, if not impossible, to avoid conflict, which is damaging to the performance of the organisation.

At first sight the objective of a construction project management system does not seem to be too difficult to visualise and in some ways it is probably easier to establish than for many business organisations. The system's objective is typified by the 'client's brief', in which the client states what he expects from the finished project. However, many clients' briefs are unsatisfactory. Often the client's requirements for cost and time for completion are not stated clearly or are incompatible, and sometimes the functional and aesthetic needs of the building are not fully or properly established. This may be caused by uncertainty created by the conditions in the project's environment, in which case the system has to respond by attempting to find ways of coping with uncertainty. It may be through lack of skill or attention in developing the brief, in which case the project management process has been deficient. In either case, it is of paramount importance that the state of development of the objective be known and understood by all the contributors to the project. In the former case, they will be aware of the degree of uncertainty inherent in the objective and should adopt approaches and techniques that can best allow for this. In the latter case they should be allowed to respond to the brief in order that they can contribute to identifying and rectifying deficiencies, so that they work towards an objective which they believe will satisfy the client.

Although the system's objective may be relatively easy to perceive, it may be difficult to articulate. The chances of conflicting objectives arising on construction projects are quite high as a result of most projects being developed by a group of independent firms and professions. The objective of a firm may conflict with the objective of the project team, and the sentience of the different contributors may lead to conflicting interpretations of the project's objectives. The manager of the project will therefore have to set the project's objectives. He must ensure that they are accepted, understood and interpreted consistently by the contributors and must attempt to resolve any conflicts as they arise.

However, systems theory has rather more to say about objectives than this rather simple view. For instance, it considers long-term and shorter-term objectives, the latter often referred to as the goal and the former the objective, but these ideas are often more useful to an analysis of the firm rather than of the project management process. Systems theory also recognises the equifinality of open systems. This means that open systems can reach the same objective from different initial conditions and by a variety of paths. The project manager can therefore use a variety of inputs in different arrangements in the organisation of construction projects and can transfer these in various ways to achieve a satisfactory output. Thus the project management function is not necessarily to identify a rigid approach to achieving the system's objectives but is to have available a variety of approaches. This view can be extended further. Not only does an open system not adopt a rigid approach; it does not necessarily seek a rigid solution, but has a variety of satisfactory solutions which may meet its objectives.

In terms of construction project management, this concept reminds us that the satisfaction of the objectives of the client does not have to be achieved by the construction of a project. A variety of solutions are open to the client. A client may decide that rather than building it may be possible to take over another company in order to achieve the objective, or alternatively choose to reorganise its own activities to achieve what is required. Even if it is decided that a building is necessary, the project manager can achieve this for the client in a number of ways, each requiring different inputs and achieving the same or different outcomes, all of which may satisfy the client's objectives. If, for example, the project manager satisfies the client by leasing an existing building, the project manager will use different inputs than if a new building were constructed. If the decision is to construct a new building, there are various arrangements that can be used to provide it, e.g. conventional arrangements, design-and-build, etc. This leads to the recognition that the client is part of the project management system or, more constructively, that the construction project management system is temporarily a sub-system of the client's organisation system for the duration of the project.

Bennett (1991) points out that due to equifinality it is impossible, except for the simplest system, to identify the route which optimises the performance of the system and its sub-systems. Hence, project managers cannot hope to optimise the performance of construction project organisations. Project managers therefore satisfice, that is they accept a solution which is sufficient to satisfy the minimum criteria for acceptance. He concludes that project managers have a continuing responsibility to search for better ways of doing things but should do so cautiously because systems as complex as construction projects cannot be fully understood and therefore the effect of changes cannot readily be predicted.

Environment

A system's environment consists of all elements outside the system that can affect the system's state, as defined in the previous chapter. This means that environments can be very complex, yet it is not possible to understand an organisation as an open system without a constant study of the forces that impinge upon it. One of the major jobs of the manager of a construction project is to relate the project to its environment. The manager should not be concerned only with the internal regulation of the system. As the system has to respond to changes in its environment, the project manager must be able to detect and analyse such changes if he is to adapt the internal organisation of the system in response to them. Project managers will be closely involved with issues and problems within the project system but their actions should be orientated to their understanding of the external influences acting upon the project organisation.

Mintzberg (1989) succinctly describes some basic organisational responses to different states of the environment of organisations generally:

'● The more dynamic an organization's environment, the more organic its structure.
● The more complex an organization's environment, the more decentralized its structure.
● The more diversified an organization's markets, the greater the propensity to split it into market-based units, or divisions, given favorable economies of scale.
● Extreme hostility in its environment drives any organization to centralize its structure temporarily.'

Understanding these types of responses in a construction setting is essential for project managers.

The process of providing a project is a response to the actions of the environment. The environment acts in two ways upon the process: indirectly upon the activities of the client of an individual project and directly upon the process itself. At its root it is the action of forces in the environment of the client's organisation that triggers the need for construction work. That is to say, the client's organisation has to respond to certain environmental forces to survive, or to take an opportunity to expand, as a result requiring construction work to be undertaken and therefore providing the construction process with work. It may be that new legislation is enacted, which means that a client's present buildings will not conform, e.g. additional requirements for fire safety in hotel buildings; perhaps the client has developed a new production process to compete with his competitors and a new building is needed to house it; a public authority may be required by law to provide a certain new service which requires new buildings. On an international level, internal or external political pressures may mean that a regime has to provide better housing or infrastructure development. In all these examples the need to construct resulted from events outside the client's organisation (or system).

The environment is fundamental not only to triggering the start of the process but also to what takes place within the process of construction. At a strategic level it will determine how the building should be provided. For instance, the state of the property market may have an important effect upon whether a building is leased or a new building is constructed. Such a decision will, of course, also depend upon the process to be housed and whether it requires a new building or can be housed in an existing property. The technology of the process is likely to determine this and will to a large part be dependent upon technological advances in the environment of the client's organisation, for example recently developed materials and machines. Similarly, changes to the proposed building required by the client during design or construction will normally come about in response to environmental forces acting upon the client's organisation.

The environmental forces acting directly on the design and construction process can affect the ability of the process to achieve what the client wants. For example, high economic activity can produce a high level of demand on the construction industry, resulting in shortages of materials, which may delay the project; industrial action can produce labour shortages; high level and uncertainty of inflation can make estimating and cost control difficult, resulting in overspending.

International projects invariably have extremely complex environments. Not only do the environments generated by the countries in which the projects are being constructed affect them, but also the environments of the countries providing the construction team. The effects can be much more pronounced than for locally produced projects for local clients. These influences are reflected in the instability of many developing countries, for example the economic, political and legal environment in China during modernisation. The action of these forces is also often reflected in material shortages in countries that do not have indigenous material availability and lack control over such supplies.

Winch (1989) draws a distinction between the impact of environmental forces on construction firms and those on construction projects. He believes that the effect of economic forces, market complexity and technological change on construction firms are not strong. Rather he believes that environmental uncertainty arises from the project's environment and the way in which projects are awarded. Such environmental uncertainty, he believes, arises from task uncertainty due to the bespoke nature of construction, natural uncertainty such as the weather and geological uncertainty and organisational uncertainty due to temporary project coalitions. He then identifies a separate source of uncertainty which he terms contracting uncertainty. He sees contracting uncertainty as due to estimating not being an exact science and small changes in the tender success rate leading to large changes in levels of turnover. Contracting uncertainty is certainty not a project environmental factor but can be perceived as one in the construction firm's environment which is strongly linked to economic and market forces which Winch does not believe are significant. Construction firms themselves may dispute the view that their environments are not particularly uncertain. It is important to recognise that uncertainty in construction firms' environments and clients' environments impact strongly on projects to compound the effect of projects' own environments.

Internationally, the impact of environmental forces tends to be much more severe than Winch indicates, presumably for UK projects. Youker (1992) in a review of the results of project monitoring and evaluation on World Bank projects indicates that many of the key problems of implementation lie in the general environment of the project and are not under the direct control of the project manager. He cites the following environmental problems as examples:

'• A shortage of local counterpart funds (the government treasury does not have the money that it promised to finance local expenditures such as the purchase of land).
• The inability to hire and retain qualified human resources, especially managerial and technical personnel (government personnel policies and procedures do not mesh with the needs of a temporary project).
• The ineffective transfer of technology and difficulty in building and sustaining institutional capacity.
• Difficulty in changing the policy environment, for example in pricing.
• Inadequate accounting, financial–management systems and auditing.
• A shortage of supplies and materials due to overall economic problems.'

If anyone should doubt the impact of both macro and micro environmental forces on projects they can do no better than refer to Morris's (1994) splendid case histories of famous and infamous major projects which, although mainly drawn from the defence and aerospace industries, also include construction.

Environmental forces

An example of how environmental forces may be classified into general groupings is shown in Fig. 4.1. These forces are applicable to any system of organisation, and may be interdependent, as illustrated. It is the interdependency of

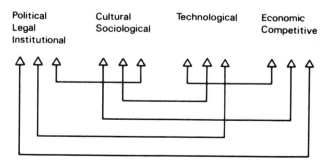

Fig. 4.1 The interdependency of environmental forces.

environmental forces that creates complex environments and makes analysis difficult.

A system receives information, energy and material from its environment, transforms them and returns them as output to the environment. Information is received, for example, regarding the economic climate and the opportunities it presents, new technological advances, the skills of people available to the system and the attitudes of trade unions and employers' associations. Energy is received, for instance, through power to drive machines and provide heat, and through computing power, but perhaps more importantly for the construction system through ideas and people imported into organisations. Material is the raw or partly or fully formed material used by the system, not only building materials but also those consumed by management and administrative processes.

The output of the construction process is returned to the environment. The effect of this can be visualised, for example, as the use to which the client puts the building and the effect on the community of the establishment of the building in a particular location, and, for commercial clients, the effects of enhanced activities on competitors and the economic climate.

The forces provide their input to the system in a variety of ways, as shown, for example, in Fig. 4.2. Environmental forces can be classified in a variety of ways and can be identified and analysed for individual projects. From such an analysis the impact of the forces and their input to the system can be anticipated. This approach will give the manager of the project the best chance of coping with them, although it must be recognised that for many projects it is not possible to mitigate the effect of all environmental forces.

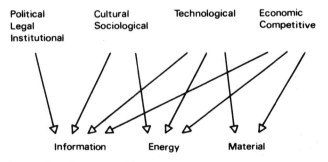

Fig. 4.2 The input of environmental forces.

The relative importance of the various environmental forces and their impact upon the client's organisation and the process of construction will vary between different classes of client and project. However, the same classes of environmental force will be acting upon each system and can be broadly visualised through the following examples.

Political

By political forces is meant the influence of government policy, for example control of the level of economic activity through investment and taxation policies, and of the distribution of activity through investment incentives. Political forces influence the availability of finance and exert effects on the labour market. Educational policies are another political force. In countries with unstable regimes, international projects are particularly sensitive to political forces.

Legal

Legislation can affect the client's activities by acting directly on the process of construction (for example, through regulations governing building, safety and planning), or by influencing the incentive to build (for example, by controlling the availability of land). In addition, legislation can affect the relationship of participants (for example, through control of monopolistic activity).

Statutory legislation is the result of political activity, but common law provisions also often have an important effect, particularly overseas. For example, Islamic law based upon the Koran has an important effect upon how disputes in building contracts are settled in some countries.

Institutional

Institutional forces include the influence of professional institutions upon the activity of their members through rules of conduct, education, and conditions of engagement. Trade and employer associations can exert effects on the activities of their members. The influences of the parent company, head office and shareholders are also institutional forces.

An inverse of these effects overseas is the forces acting upon the contributors to projects in which such institutional constraints do not exist, and their need to respond competitively in such circumstances.

Cultural and sociological

The acceptability of specific activities by the general public, particularly as reflected by the local community, is an example of a cultural and sociological force. The effect of events in the world on the values and expectations of employees is another example of this type of force, as is the influence of trade unions and of informal contacts upon members of the system. An example of a significant effect of such forces overseas is that, when building within the walls of the city of Mecca, only Muslims can be employed in any capacity on the site.

Technological

Technological forces include the influence of technology on processes through the development of new materials, techniques and ideas and through the experience of others with those materials, techniques and ideas. The current development of technology and its potential for solving problems is an obvious example of a technological force. An extremely important influence in this category is the massive increase in cheap computing power now available.

Economic and competitive

Economic and competitive forces include the level of general economic activity and the demands this places upon organisations. The state of competition, the effect of monopolistic phenomena, the availability of finance, materials and labour, and the level of interest rates are other examples of economic and competitive forces. This category is, of course, very closely related to, and dependent upon political forces.

Other approaches

Two developments of the contingency approach have focused on the effect of open systems of specific environmental components in terms of how they influence organisation structures. Institutional theory focuses on institutional forces in the environment. The interpretation of 'institutional' is broader than that given above. Greening and Gray (1994) state that:

> 'Institutions specify rules, procedures, and structures for organizations as a condition of giving legitimacy and support (Meyer & Rowan 1977). These institutions have traditionally included state and federal governments and professional groups as well as interest groups and public opinion (DiMaggio & Powell 1983; Tolbert & Zucker 1983). Also that institutional theory seems particularly well suited to explaining the development of issue management structures because issue management focuses on external influences generated within the public arena.'

Such forces are becoming increasingly relevant to construction, for example in the light of increasing environmental concerns both in terms of pollution and the potential effect of construction work on ecology.

The focus of resource dependency theory emphasises that firms adapt their structures in the face of dependencies on external organisations, and hence managers can exercise some choice over organisation structure within the constraints of the environment. The theory believes that organisations are constrained by organisations (institutions) in their environment upon which they depend for critical resources and that hence organisations seek to manage their dependencies on external organisations to gain more authority and freedom. It can be seen that resource dependency theory has elements in common with institution theory but that its dependencies are more widely drawn. It also draws from the transaction cost framework on how organisations are formed. In exercising choice within the constraints of the environment, managers will seek to manage their dependency on external organisations. This is illustrated in construction by construction firms' decisions on whether to subcontract or carry out work in-house and whether to own or hire plant.

In developing these theories, elements across all the classes of environmental forces previously identified are incorporated which further emphasises the causal texture of environments (Emery & Trist 1965).

Action of environmental forces

The interaction of environmental forces and their consequent effects on the client and the construction process determine the climate in which the system exists. A low level of activity of environmental forces upon a system will lead to a relatively stable system, whereas a high level of activity will lead to the system existing in an uncertain climate.

In terms of the construction process, environmental forces act in two ways:

(a) upon the client's activities and hence transmitted to the construction process (*indirect*);
(b) directly upon the construction process (*direct*).

The process exists, therefore, in a complex environment, as illustrated in Fig. 4.3, which must be reconciled in the interests of the client. In circumstances where the indirect and direct environmental influences act in a conflicting manner, the project management process will be required to attempt to resolve the conflict to the benefit of the client. For example, the contractor may wish to move labour from the site to aid the profitability of another contract (an influence acting directly on the construction process), which may put at risk the completion of the

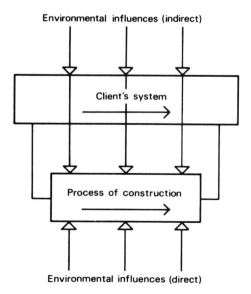

Fig. 4.3   The environment of the process of construction.

building on time when the client's environment demands completion on time. In such a case it is the duty of the manager of the project to resolve the issue in favour of the client.

The environmental influences acting directly upon the client's organisation should determine the organisation structure and mode of operation appropriate to the client's activities. In addition, environmental influences will present opportunities to the client, and will determine the manner in which such opportunities need to be taken. For example, a client's environment may determine that an additional manufacturing capacity needs a building quickly in order to take advantage of an opportunity. In such circumstances, the organisation set up to achieve this must be capable of acting quickly. If, at the same time, forces indicate that uncertainty of the size of the market for the goods is likely, the organisation set up to take advantage of the situation must also be capable of achieving the flexibility required. However, an unexpected large order for the client's goods may make the need for a new building urgent, but it may occur at the time of a rise in activity in the building industry. This may create uncompetitive conditions in terms of price and completion time for projects and make it difficult to achieve completion when required. Project management must strive to overcome this type of problem caused by conflicting environments.

The construction process is therefore made complex by the type of environment in which it exists, which creates a need for high-level managerial skills. The process must produce a clearly defined solution at the technical level of design and construction but must also remain flexible and adaptive to satisfy environmental requirements. The managing system will be required to reconcile these competing demands, which become more difficult as environmental complexity increases and in many cases may be incompatible.

The ideas developed here see the process of construction as a sub-system of the client's system. As such, it is influenced by the client's environment as well as by the particular environment of the process. This is a development of the tentative view of the Tavistock Institute (1966) which, although not conceiving the process of construction as a sub-system of the client's system, drew attention to the obsolescent nature of the concept of the architect 'taking a brief from the client' in the conventional way.

Recognition of the construction process as a sub-system of the client's system identifies a boundary between the process and the client's organisation that needs to be integrated. The need for integration has as great an implication for the client as it has for the construction team, since it will demand that both systems establish appropriate ways of achieving the level and style of integration demanded. Both a National Economic Development Office (NEDO 1975) report and the Tavistock study referred to the lack of such integration in the conventional process and the development of project management skills continues to need to concentrate upon this aspect today.

The implication of the relationship between the systems is that changes in the elements of a client's environment or their relative properties may require a change in how the project is designed, constructed or procured. This may happen during design or even during construction. The integrative device at the boundary between the client's system and the construction system should recognise and take action on changes in the client or construction process environments in terms of maximising the benefit or minimising the deficit to the client. This should be the objective of the integrative device.

The relative uncertainty of environments and the nature of the tasks of both the client's system and the construction system should determine the nature of the integrating device and the organisation structure of the construction process. For example, in an environment that is economically or technologically uncertain or both, the organisation structure of the process should be designed to be sufficiently organic to respond to stimuli. This should be reflected in the style of integration used, for example by the project manager assuming a predominant role. This necessity can be visualised, for example, in large-scale long-programme hospital development. Conversely, a stable environment could more readily accept a more mechanistic organisational structure, with integration based on standard procedures and routines, for example in small school building.

Many of the features discussed here are illustrated by Walker and Kalinowski (1994) including the effects of strong macro environmental effects and the manner in which the organisation structure was designed to cope with them. The findings are incorporated in the fourth case study in Chapter 13.

Measurement of environmental uncertainty

Whilst the ideas of environmental uncertainty can be readily conceptualised, its formal measurement has received little attention by researchers (Buchko 1994) no doubt due to the sheer complexity of environments and hence the difficulty of devising convincing methodologies. Buchko provides a good review of research in this area which begins with Lawrence and Lorsch's original work on uncertainty associated with a specific job in an organisation. Researchers moved this on to applications to the firm but were faced with increasing complexity.

Buchko believes that:

'The conceptualizations of the environmental uncertainty construct used are not consistent, ranging from predictability through dynamism and complexity to controllability. Such differences in the underlying conceptual definitions of the environmental uncertainty construct make generalizations or interpretation of results difficult.

Several studies have used measures of perceived environmental uncertainty that are idiosyncratic to the particular research effort and that their conceptual definitions of uncertainty have varied significantly; such concepts as turbulence, complexity, predictability, and heterogeneity have all been included, making interpretation of results across studies problematic.'

As a result of his own research using the Miles and Snow Perceived Environmental Uncertainty Scale (Miles & Snow 1978) Buchko concluded that:

> 'Perceptions of uncertainty may be inherently unstable because environmental complexity and dynamism may prevent individuals in organizations from developing stable assessments of the environment itself. Organizational and individual characteristics may affect the perceptual process ... yet the difficulties posed by the natures of organizational environments and managerial perceptual processes may make the development of reliable and valid measures very difficult.'

Hughes (1989) used a much less ambitious approach in his attempt to describe rather than measure the environments of construction projects. Using eleven environmental factors, e.g. economic, cultural, etc., he scored each on a scale of variability of 0, 1 and 2 for each of five variables – definition, stability, certainty, simplicity and mitigability. A problem with this approach is that the variables may not be mutually exclusive. Nevertheless the approach is useful for helping managers understand the environments of their projects, and for giving discipline to project scanning. Project scanning is the process of analysing the project environment to define potential problems and assessing the probability of their occurrence. Youker (1992) believes that there are several different ways of scanning the environment each of which contain the categories of environmental forces identified previously:

'• Elements that are suppliers of inputs, consumers of outputs, competitors, and regulators.
• Elements that are physical (e.g. climate), infrastructural (e.g. power supply), technological (e.g. plant genetics), commercial/financial/economic (e.g. banks), psychological/sociocultural (e.g. attitudes toward credit risk), or political/legal (e.g. local government).
• Elements that are hierarchical and sometimes geographical, such as government at various levels, i.e. national, regional and local.
• Elements that are actors (e.g. individuals, groups, institutions) or factors (e.g. attitudes, trends, laws).'

Whilst these illustrations are drawn from World Bank projects, many will be familiar to those involved in construction projects. Youker points out, very practically, that analysis of the key elements of the environment will not solve all problems as some will be intractable but at a minimum will give early warning of problems so that the best chance of finding a solution will be provided.

There is therefore no easy or precise method of quantitatively assessing environmental forces and their impact on construction projects. Indeed, if there were, it would be a large step towards solving the problems faced by governments let alone by construction teams! Nevertheless, recognition of environmental forces and an understanding of how they affect construction projects does allow those involved in their management to construct a scenario for their work which should allow them to anticipate and respond to changes in environmental conditions. Preconceived methods of working all too often have fixed rigid patterns for project achievement. This can mean that the process is unable or unwilling to respond to environmental changes to such an extent that although a client's objective may have changed the objective of the construction process has not.

A construction project organisation should therefore be designed to reflect the type of environment in which it is to work. There is little point in using a well tried rigid organisation structure suitable for building public housing for the design and construction of a second Sydney Opera House. The conventional UK approach to organisation of the construction process is not directly transportable to overseas projects for similar reasons. The environmental conditions in which such projects are designed and built are very different from those in which the convention was developed.

Negative entropy, adaption and protected environments

Entropy is applicable to all closed physical systems. It is the tendency for any closed system towards a chaotic or disorganised state in which there is no potential for further work (Katz & Kahn 1978). This may sound rather dramatic for a book about construction project organisation but it has some significance. Construction project organisations are open systems and open systems attempt to find ways of avoiding such a fate. They develop negative entropy (negentropy), which is a process, or more complete organisation and more ability to transform resources. They achieve this by importing the resources (material, energy, information) from the system's environment. Social organisations such as those involved in construction can continue to import new human and other resources to allow them to continue functioning and may be capable of indefinitely offsetting the entropic process in a way which open biological systems and closed physical systems cannot.

In developing negentropy, an open system may be seeking to achieve a steady state in which the system remains in dynamic equilibrium through the import of resources from its environment. That is, it attains stability or is self-regulating. This view is more relevant to biological systems and allows them to cope with variations in their environment. For example, the human body can maintain a steady state in spite of wide variations in the environment. There are of course limits; environmental changes may be so great that the system dies.

Taking these ideas a stage further introduces the adaptive system (Buckley 1968). A system is adaptive when it changes its own state and/or its environment when there is a change in its environment and/or in its internal state that reduces its efficiency. Adaptation is therefore the ability of a system to modify itself or its environment when either has changed to the system's disadvantage. Complex adaptive open systems allow interchanges among their internal components (or sub-systems) in response to environmental forces to such an extent that the components themselves may change and hence the system as a whole may adapt and so survive.

It is interesting to assess whether the construction process fits into this systems scenario and whether it helps our understanding. The firms that contribute to the construction process do import material from their environments in the form of new staff, new ideas, new technology, etc., and so develop negative entropy. However, construction project organisations are generally temporary. They cease on completion of the project and further organisations will be formed with either the same amalgam of contributors or with different partners. The process itself is not therefore truly negentropic except in the case of design-and-build or turnkey organisations. The firms themselves will, however, be importing from their environments so that they are in a position to be able to join future project organisations. In the case of design-and-build and turnkey organisations, one firm is responsible for practically the whole process and in these cases the firm and the process attempt to develop negentropy.

In adapting to their environment, some systems will attempt to cope with external forces by acquiring control over them. This process can be seen in the mergers of companies, often to reduce competitiveness in their environments, which result in the expansion of the original system. Some organisations may have achieved such a degree of monopoly or have acquired a protected niche in the environment to the extent that they can ignore a certain level of environmental pressure. If this occurs, such organisations can afford to accept a level of suboptimal performance and can survive at that level.

In the case of the construction industry's professional and industrial firms, for many years the amount of adaptation to their environment was not great, as illustrated by the large proportion of projects undertaken predominantly on the conventional pattern in spite of much criticism of this process. The conventional pattern of organisation tended to be self-regulatory and to function to maintain the given structure of the system. This was due, to a large extent, to the system existing in an environment from which it had protected itself. This was achieved through codes of conduct and fee scales of its professional institutions, which eliminated, to a large extent, competition between firms, thus enabling the system to resist change and maintain the status quo.

However, the increasingly competitive environment in which the system and its clients have to exist have been significant in breaking down such protection. In the case of clients they have brought to bear greater pressure for change in the industry's procedures as a result of the increased competition with which they themselves have been faced. The increasingly multinational nature of the industry's clients and overseas practice have also been major forces for change as clients have experienced methods of managing the construction process that differ from those used conventionally in the UK.

Although many of the professions' and industry's firms try to protect themselves at an institutional level, changes are taking place at the project level with the increasing use of such processes as USA-style construction management, design-and-build and build-operate-transfer. At the project level it appears that some firms are adapting by changing the nature of the internal components of the building process, for example by introducing the contractor into the design team, and so moving nearer to the open adaptive system described earlier.

In view of the traditional institutional domination of the professions' and industry's firms, it was always likely that adaption would take place at the project level in response to the demands of clients. However, this was only likely to take place for projects with clients who themselves were adaptive and not protected in some way from their own environments. It is not surprising, therefore, that many initiatives that have taken place in the management of construction projects have been for private commercial or industrial clients, and that there was a lack of stimulus from public clients who were themselves protected to a large degree from their environments. However, with the change in local government culture in the UK changes also took place in this sector's projects. The process of providing a construction project should be an open adaptive system but it may be constrained by the environment within which it exists. Nevertheless, the process needs to change its structure if environmental events, acting either directly upon the process or indirectly upon the client's organisation, dictate that this should happen.

Growth, differentiation, interdependency and integration

Whereas closed systems move towards disorganisation (Kast & Rosenzweig 1985), open systems move in the opposite direction towards a higher level of organisation, which generates greater differentiation among their parts (sub-systems). This feature is observable in business organisation systems and can occur in two ways. One, which has been referred to previously, is when a system seeks to encompass parts of its environment and annexes them as sub-systems, for instance when one firm acquires control of another. The second way by which it occurs is where complex and uncertain environments create the need for sub-systems to specialise further in order to cope with such complexity and uncertainty. That is, the level of skill required is such that a sub-system cannot cope with the range of skills demanded of it and it has to subdivide further. In traditional management thinking, specialisation was considered to be 'a good thing' for increasing the efficiency of undertaking a particular task. Differentiation, however, is now explicitly considered to be necessary in order to allow each sub-system to cope effectively with the part of the system's environment which is acting upon it. Hence, open business organisation systems tend to grow by expansion and by internal elaboration. This is not to say that this is necessarily a benefit in all cases but simply that it is a feature of open systems. It brings along with it the greater management problems of handling large complex systems and hence the need for careful organisation design if such systems are to be effective.

The protected environment of the construction process limited the impact of these concepts upon it. However, in recent years there have been more examples of growth through amalgamation of professional firms, creation of design-and-build companies and client organisations acquiring an in-house capacity for designing and constructing projects by taking designers and construction workers on to their payroll. Similarly, a number of consortium firms including architects, quantity surveyors and engineers have been formed. All of these arrangements helped organisations to handle more easily the environment in which they operated. A growing area of such activity is the joint-venture movement in which firms form separate joint companies to bid for and undertake projects. Companies make such ad hoc arrangements between different types of firm, e.g. architects and contractors, and between similar types of firm, e.g. contractors. In most cases the aim is to spread the risk in the difficult and uncertain environmental conditions that usually accompany large and complex projects, which are often overseas. Such activities are also predicted by the transaction costs framework, which indicates that management theories are not mutually exclusive.

Growth through internal elaboration has occurred relatively slowly in the construction industry. After many relatively stable years, specialisation into architect, quantity surveyor, specialist engineers and contractors took place quite quickly during the last century and the early 1900s. This resulted from the increasing complexity of the environment in which construction took place during industrialisation. The process then slowed down as the professions protected themselves from their environment and attempted to maintain the status quo. The subsequent proliferation of specialist subcontractors can be seen to be further differentiation to cope with complexity and uncertainty. There is also evidence of the same phenomenon in the specialisation of quantity surveyors into tender document production, construction economics and cost control. Similarly, architects often specialise in particular building types. Once again such movements reflect the transaction cost framework as well as systems theory in a manner which is the converse to that described above.

The notion of interdependency is explicit in the earlier definition of a system, that is, it is an entity consisting of interdependent parts. If this is the case, then the greater the differentiation of the interdependent parts of a system, the greater will be the need for integration. Differentiation in organisations has been defined (Dalton *et al.* 1970) as the differences in cognitive and emotional orientations among managers in different functional departments and the differences in formal structure among these departments. Integration has been defined (Dalton *et al.* 1970) as the quality of the state of collaboration that exists among departments that are required to achieve unity of effort by the environment.

The interdependency of the contributors to the construction process has long been recognised but often as sequential interdependency. In other words, one part cannot act until after the previous part has done its work. It has become increasingly recognised that in fact interdependency should be reciprocal, that is iterative, and the process should move forwards following decisions to which all appropriate parts of the system have made a contribution. The integration of reciprocal interdependencies requires considerably more skill and effort than the integration of sequential interdependencies. It is the recognition of this fact that has focused attention upon the management needs of projects. Recognition of the need for project management has been highlighted by the complexity and uncertainty of the environment within which construction takes place, which as led to greater differentiation within the construction process and hence to a greater need for skill and effort in integration.

At the extreme of the large complex overseas project it is not unusual to find a number of 'separate' projects going on at the same time using a range of architecture, quantity surveying, engineering and other consultancy firms and a variety of contractors and subcontractors all working on the same site for the same client who has an overall objective for the development. It is not difficult to visualise the integrating effort required in these circumstances. Even at the 'smaller' end of the market, the rehabilitation of an area of a city involves numerous firms all working for a common client with one set of objectives.

Feedback

The concept of feedback is fundamental to understanding how a system is maintained and therefore how it continues to exist and accomplish its purpose. Feedback is the basis of a systems control function. It is through feedback and subsequent action that achieved outcome can be compared with desired outcome so that adjustments in the behaviour of the system can be made. The need for a control function for construction projects is self-evident and much of the energy expended in developing techniques in recent years has been directed at achieving more sophisticated control. However, the type and the amount of feedback designed into a system are the key to the system's stability and economy and in this respect it is interesting to note that the control mechanisms on construction projects are often no more than monitoring devices that declare the position too late after the event to take corrective action, e.g. many 'cost control' procedures. Feedback points should be carefully designed into the system so that appropriate action can be taken at the right time. Feedback should operate on a cost-effective basis in such a way that the value of the control achieved is not cancelled out by the cost of achieving it.

The operation of a feedback loop requires a sample taken at specifically designed points of the system's operation (often referred to as freeze points) to be measured against the objective of the system. For construction projects the sample points need to be chosen on the basis of the nature of the actual project and its environment. That is, for a simple project in a stable environment it is to be expected that only a small number of sample points will be necessary, whereas for a complex project in an uncertain environment, frequent sampling will be required. Naturally, this means that the objective of the system should be appropriately, accurately and explicitly defined to enable the control mechanism to carry out its function. It is questionable whether many of the 'client's briefs' commonly used in the construction industry are sufficiently clear to allow this to happen. Effective control systems require that the procedure for testing the sample against the objective be designed with appropriate methods of measurement of the sample against the objective and, importantly, with the ability to take action on the basis of feedback information.

The conventional organisational structure of the construction process often does not possess this ability as the relationships of the contributors to the process are arranged in such a way that the people reporting on the current state of the project *vis à vis* its objective are not in a position of sufficient authority to ensure that the project returns to its intended course. This is often the case when the architect is in both an operational capacity as a designer and also in the primary management position for the project. The application of a systems approach to the design of organisations should automatically establish relationships which would allow a properly designed control function with appropriate feedback mechanisms to overcome this deficiency and operate effectively.

The simplest kind of feedback is negative feedback. This enables the control function to correct the system's deviation from its course, that is, it encourages a return to the initial objective. Most control functions used on construction projects operate in this way by attempting to correct deviation in cost, time or design of the project and return it to what was intended by the 'client's brief'.

Positive feedback, on the other hand, further amplifies deviation from course, as a result of redefinition of the system's objectives. Although this may be an unusual reaction for construction projects, it should not be overlooked. If, during design, it is discovered that the provision of facilities in the project is deviating from what was intended, it may be that the client's original requirements have changed, and upon seeing the developing design the client may decide to continue along this course. The objectives would therefore require to be amended in response to positive feedback. Such change could be as a result of a badly constructed brief but could also be caused by changes in the environment of the client's organisation that have altered their objectives. The control function should therefore operate within the system and between the system and its environment.

The nature of the process of designing and constructing is characterised by a series of 'pinch points' through which it must pass if progress is to be made. At each pinch point a decision has to be made, for example whether the design satisfies the function required of it, whether the cost is acceptable, whether the proposed procurement method will allow the project to be completed on time, etc. The decision points can be conceived as a hierarchy with decisions taken by the client at the top, those taken by the manager of the project at the next level, and those taken by the operational people at the lowest level. The decision structure of a project can be used to provide the control framework. Feedback can take place at each decision point to test whether the proposed decision will help to achieve the objective of the total system. It should be noted that, on many projects, decisions are not made explicit and therefore are not used in this manner. They are not consciously tested but are assumed to be correct as 'that is the way we have always done it'. However, by anticipating decision points and the nature of the decisions to be taken, a control framework can be established and the contribution to be made by each participant can be designed using systems principles.

Summary

The process of designing and constructing a project on behalf of a client can be analysed as an open adaptive system. As such it needs to respond to its environment but historically it has, to a degree, protected itself from its environment by the construction of rules, procedures and conventions which have been granted validity by public authorities, professional institutions and other bodies associated with construction. Nevertheless, the environmental influences upon the process, particularly those being transmitted to it through its clients, have resulted in the process becoming more responsive.

Analysis in systems terms focuses attention upon the need to bind together the differentiated yet interdependent contributors to the process. This requires a high level of integrative activity which has not traditionally been recognised and provided. The provision of integration must be directed towards the achievement of the total systems objective, which must be stated unambiguously in terms to be the client's requirements.

The determining factors of how the system is structured and operates to achieve its objectives are the technical demands of the project, together with the environment in which it is undertaken. The control function should therefore be designed to reflect these factors and be based upon the anticipated decision points in the process. The decision points will determine the interdependency of the contributors to each decision. Therefore their relationships should be designed on systems principles in terms of their contribution to each decision. Such an analysis demands that the organisational structure established for each project should be developed individually from first principles, and although a range of 'standard solutions' may emerge, it should not be presupposed that any one solution is automatically the correct answer.

Returning, briefly, to the environment and its implications for the construction industry and its professions, it has been observed that industrial companies often fail to obtain adequate information about changes in environmental forces and that it is remarkable how weak they are in their market research departments when they are so dependent upon their market. It has been predicted that organisations will improve their facilities for assessing environmental forces and hence improve their marketing ability. The same could be said of the construction industry and its professions now, and perhaps the same prediction can be made.

The remainder of this book is developed against the background of this systems scenario of the construction process, which can give structure to our understanding of the process. The systems view clarifies how the process works in practice at the present time and points the way to how it may improve in the future to enable it to satisfy more effectively the requirements of its clients.

Chapter 5
The Client

Introduction

On the face of it the use of the term 'the client' in the construction industry is simple enough but its apparent simplicity hides a complex concept. As early as the mid-1960s the Tavistock Institute (1966) was drawing attention to the increasingly intricate nature of client organisations, saying:

> 'that they were complex systems of differing interests and that their relationship is seldom with a single member of the building industry ... These client systems are made up of both congruent and competing sets of understandings, values and objectives. Much design and even building work has proved to be abortive because unresolved or unrecognised conflicts of interests or objectives within the client system have only come to light after the building process has been initiated'.

Since that time clients have been putting their views more forcefully. For example by Mobbs (1976) of Slough Estates who accused the UK construction industry of poor performance and subsequently of failing to satisfy the needs of clients, and particularly purposefully by the British Property Federation (1983) which devised their own system for commissioning design and construction due to their dissatisfaction with conventional methods at that time. Increasingly since then, the professions and industry have responded to clients' criticism. Clients have, therefore, had a great influence on the industry and have induced changes in attitude which have seen the industry adopting a wider range of organisation forms and greater flexibility in seeking to meet clients' objectives.

Where the client is to be both owner and occupier of a building, the idea of a client appears straightforward but even in such circumstances may not be so. More frequently the client will not be both owner and occupier and the situation may be more complex. For example, the owner in the first instance may be a property developer who sells the property to an investment company which then leases it to an occupier. In the public sector the client may be a local authority which receives finance for the project from central government to implement government policy, and the users may then be a third party, for instance teachers and pupils in the case of a school. There are many more similarly complex cases and thus the first question that needs to be answered is: who is the client?

A construction project team will tend to recognise its client as the body that has the authority to approve expenditure on the project, the form that the project has to take, and its timing (and who pays the fees). The project team will find it simpler if all these authorities are vested in one body but frequently this is not so in practice. For example, central government may allocate funds for a project to a local authority, which will be responsible for developing the project, but it may reserve the right of final approval of both expenditure and aspects of design. A similar situation can occur between head office and regional office of a private client. If another group receives or occupies the building after completion, it will be the client of the client of the construction process, and the responsibility for satisfying that client will rest with the commissioner of the project. Nevertheless, in some instances, the project's client may wish to involve such a third party in approval of the design and this can further complicate identification of the client for the project team.

The members of the project team need to have the ability to understand the structure of their client's organisation, their relationship to others with an interest in the project and what makes them tick. In particular they should understand the decision-making mechanism of the client's organisation and where authority for decision lies. Only when this is known will the project team be in a position to obtain information upon which it can act with confidence.

Having understood the organisation structure of their client and how it operates, the project team will be in a position to build up the trust and confidence necessary for it to obtain accurate and useful information that will enable it effectively to develop an appropriate brief for the proposed project. Bearing in mind that every company and public authority and even every individual is potentially a client of the construction industry, the breadth of knowledge of organisations required by members of project teams is so vast as to be unrealistic. Project teams therefore need to acquire some conceptual tools such as those discussed in Chapter 3 with which to analyse and understand their clients' organisations. An indication of the scope of the construction industry and hence the wide range of private and public clients with which it has to deal is given by Table 5.1.

However, the reality is far from such a complete understanding of client organisations on the part of project team members. It is more frequently the case that members of the project team do not meet the client. The project team leader acts as surrogate client in many traditional contracts (National Economic Development Office 1975) and the team members know the client's requirements at second and third hand only.

Clients also have important responsibilities to fulfil which cannot be delegated to the project team. The National Economic Development Office (1975) emphasised, in respect to public sector clients, the strategic role of the client in the area of selection of project team members, setting key dates, brief development and

Table 5.1   Distribution of contractors' new orders, Great Britain, 1994. (*Source:* Department of the Environment 1995).

	Public sector	Private sector	% of all new orders
	Current prices (£m)		
New housing	1 386	5 721	33.4
Infrastructure	2 211	1 240	10.3
Factories and warehouses	149	1 949	9.8
Oil, steel, coal	12	51	0.3
Schools and colleges	658	} 115	} 5.4
Universities	376		
Health	752	255	4.7
Offices	469	1 777	10.6
Entertainment	308	928	5.8
Garages	49	300	1.6
Shops	14	1 453	6.9
Agriculture	22	120	0.7
Miscellaneous	844	127	4.6
Total	7 250	14 036	
%	34	66	100.0

monitoring of the project. Whilst the client may wish to delegate many aspects to the project team leader the warnings of the National Economic Development Office (1978) should not go unheeded. It pointed out that 'the standard of service given by the building industry relates closely to the amount of effort expended by the client in establishing a good brief' and that 'satisfaction at the construction stage is closely linked to the degree of control and supervision by the client himself'. It is therefore important for clients and project team leaders to ensure that clients are appropriately integrated into the project's organisation structure.

Such sentiments were still being echoed and found relevant by Ward *et al.* (1991) in the early 1990s when they concluded that it is important for clients to:

- '• Set clear objectives.
- • Subject objectives initially set to careful trade-off analysis.
- • Consider objectives carefully when choosing a procurement method.
- • Communicate objectives clearly to other involved parties and avoid conflicting guidance to different parties.
- • Ensure that reaction to unexpected events involves proper revision and consideration of client objectives.'

At about the same time Thompson (1991) considered that clients should be taking an even more proactive stance when concluding that:

'The owner of a project must provide clear *direction* and timely *decisions*, and must assist the project-management team to drive the project to a successful conclusion. He or she must accept the risk associated with the enterprise, and assume particular responsibility for:

- the selection of the project team,
- thorough appraisal and realism over risk,
- championing the project in the political and public arenas.'

Most importantly he pointed out that corporate client organisations are rarely suitable for providing client management of projects as the style of project management is likely to be more dynamic then that of corporate management particularly when the latter has a rigid hierarchial management structure linked to slowly changing long-term objectives. He recommends that a strong temporary corporate project team should be formed to support project management.

Classification of clients

Potential clients of the construction industry are a too large and varied group for any meaningful detailed classification to be prepared. Nevertheless an understanding of clients is aided by a broad categorisation. What follows is, initially, a simple grouping to identify the basic types of clients followed by an account of two attempts to classify clients using more sophisticated approaches.

The individual client

The individual client is the exception for most construction projects, particularly where the client is to be both owner and occupier. But even at this level the client can be more complex than expected. A particularly simple example is a couple proposing to have a house built for themselves. In such a situation there is a direct relationship between the clients and the leader of the project team, and communication of information should be straightforward. However, even at this level it can become complicated. Who is the client: the husband, who may be providing the finance, or his wife who will probably be the main user, particularly in the kitchen (or vice versa, to avoid being sexist)? Relationships between married couples vary considerably so the project team needs to understand the particular relationship, which could be difficult! This may be taking the point a bit too far, but it illustrates the problem.

A similar but amplified situation occurs in the case of the sole owner of a business. In this case the relationship between owner and employees is important. Will the owner instruct the project team alone or will the workforce also be involved? Is the owner able to indicate clearly to the project team the activities to be housed or will the workforce need to be consulted? If they need to be consulted, what does the team do if the owner is not sympathetic to their views? Even at this relatively simple level the way in which the team obtains the information it needs can depend upon understanding the client's activities, organisation and relationships.

The corporate client

The broad classification of corporate client includes all companies and firms controlled other than by a sole principal. This group therefore covers all companies from the small, simply structured organisation to the massive multinational corporation. The myriad of functions, sizes and structures of firms within this group poses particular problems for the project team. If it is to carry out its work well, the team will need to understand the objectives of the corporate client, and these will often be complex. In particular, it will need to understand the purpose of the project for which it has been commissioned and how it is intended to contribute to achieving the client's long-term objectives. To understand the objectives of the client, and to establish the firm's objectives, it will have to be familiar with how the client's organisation operates. Such knowledge is also required to identify where the best information is likely to be available on which to base the project proposal. It will also be necessary for the project team to be able to assess the ability and status of the members of the client's organisation who are transmitting the information to them.

As no two clients within this group are likely to be structured identically, the organisation analysis skill demanded of the project team is very high indeed. Coupled with this is the need for the team to be able to build up and maintain the confidence of the client, for only if this exists is the team likely to be able to obtain the information it needs to do the job effectively, much of which may be of a confidential nature.

Perhaps the only common component of such companies is that final authority will lie with the board of directors or equivalent group and in some companies it may in reality lie with only one member of the board or with a small group of directors. The leader of the project team will have to cultivate confidence at this level. Nevertheless, in the more complex company interrelationships that frequently exist, it may be that full authority does not lie with the board but with a board of another company which has control of the client's company. Such situations can make it very difficult for the project team to proceed with confidence as decisions may be overturned to the serious detriment of progress on the project. Insights into the nature of client organisations are provided by organisation theory and the work on power in organisations. For example, asking which of Mintzberg's organisation classifications fits your client organisation can be illuminating as can an understanding of the political forces at play.

Frequently the project team will have to talk to a large number of people in the client's company to tease out the brief for the project and to develop it into an acceptable final proposal. This does not have to be restricted to managers but may also include the operational people in the company. Often clients may not know clearly what they require. The briefing process requires a large number of important decisions to be made by the client and the source and authority of such decisions have to be identified. These decisions are not just about the functional attributes of the building but most importantly about the time scale and budget for the project. Of particular importance is the timing of decisions by the client on important aspects of the project, e.g. budget approvals, as the incidence of decisions will have a fundamental effect on the progress of the project.

It is not uncommon to find that the client's company will appoint a project co-ordinator or internal project team from the company to act as the link between the client and the project team. This has been found to work successfully but it does, of course, depend predominantly upon the quality of the person(s) appointed. What is of vital importance is the authority of the person(s) in this position. If members of the project team are to rely on that person's instructions, they need to be sure that they have the authority of the board to issue instructions. If they have not, then the result will probably be frustration, delay and abortive work. The ground rules need to be clearly laid down with the client's board of directors at a very early stage. This presents a very real problem as the project team is likely to be rather diffident about 'pushing too hard' for fear of offending an important client and this is often linked with a reluctance to pry into the client's affairs, although both are necessary if an acceptable project is to be realised effectively.

Dealing with a client can be a very frustrating business, particularly in the case of large clients, which tend to move towards a bureaucratic form of organisation. This can result in the project team adopting procedures that result in their designing what they think the client wants without basing it upon investigation of what the client needs. The orientation of the project team should be strongly towards finding out rather than constructing a series of cockshies for the client to criticise. This demands skills of investigation and a large measure of diplomacy.

The public client

Public clients include all the publicly owned organisations that have the authority to raise finance to commission construction work. In all such cases the funds will normally be raised by taxation or in the money market on the authority of the government. They include local authorities, government agencies and the government itself. Normally, authority to spend money on construction stems from the government but usually, when authority has been given, the 'client' may control the spending of money within certain constraints, although withdrawal of authority is not unknown.

Many of the features described above for the corporate client are applicable to public clients, particularly government agencies but the situations encountered are often more closely constrained and difficult through having to work through committees whose authority may not be clearly defined. The bureaucratic rules that surround most decisions to construct for public clients can often lead to an inefficient construction process. A common example is the establishment and approval of a budget for a project and the limitation of having to place the construction contract during a particular financial year. This approach can produce situations in which the cash flow and budget are inflexible, and so may inhibit the project team's ability to obtain value for money, particularly where virement between different expenditure headings is prohibited. Similarly, value for money may be difficult to achieve if a project has to be rushed in order to meet a financial-year deadline.

Such difficulties may have to be faced within a structure which requires that the project team is instructed by officers of the public body who are subject to control by a committee of elected or appointed representatives. They themselves may be controlled by higher level committees, either within the public body and/or within government. The process may also include the involvement of central government officials. In such circumstances the difficulty of knowing just where the decision making is done can be a severe problem.

The project team has to develop skills in understanding how such organisations work. The wide range of public clients and their objectives, many of which may be politically generated, place great demands of client analysis on the project team. Objectives can be difficult to pin down and unresolved conflicts may exist between the various client interests and the potential users of the project. If final authority in such situations rests with an elected committee, the outcome may be unpredictable and the project team will be faced with uncertainty.

As with corporate clients, there can be great benefit in working through a co-ordinator from the client's organisation provided that his authority is clearly stated and understood by both the client and the project team. The ground rules for the validity of information and decisions have to be laid down at an early stage, but even the ground rules may not be protected if there is a change in policy, often resulting from a change in composition of the elected committee who have final authority over the project.

Uncertainty and complexity stemming from the nature of the client body is compounded on overseas projects because of the project team's unfamiliarity with the client's organisation. Many of the differences arise for historical or cultural reasons. A significant amount of homework needs to be undertaken on both the influences at work in the country as a whole and on the culture and attitudes of each client.

Client profiles

An indication of the range of client types is provided by Bresnen and Haslam (1991) in their study of client attributes and project management practices. They sampled a list of clients drawn from the 'Awards' section of the *Contract Journal*. Their method of selection reflected the importance of large-scale and experienced clients. The percentage of each client type, each project type and the project status is given in Table 5.2. The statistics give a good impression of the weighting of client and project types served by the industry. Private sector clients dominated (between 1984 and 1986), representing 65 per cent of all clients with offices being

Table 5.2 Bresnen and Haslam's study (1991).

Client type

	%
Government department	1
Local authority	22
Statutory authority	7
Nationalised industry	1
Development corporation	1
Housing association	3
Property developer	20
Company	45
Other	1
Total	100

Project type

	%
Industrial (factories/warehousing)	17
Offices (industry/commerce)	28
Commercial/retail	19
Housing	17
Education/training	7
Civic	6
Health	4
Transport facilities	3
Total	100

Project status (i.e. frequency as a construction client)

	%
First ever	12
Few before	14
Regular small number	22
Large number	43
Entire workload	9
Total	100

(*Source:* Bresnen, M.J. & Haslam, C.D. (1991) Construction industry clients: a survey of their attributes and project management practices. *Construction Management and Economics*, **9**: Table 2, p. 330 & Table 3, p. 331.)

the largest project types followed by industrial, commercial and housing in about equal numbers. The project status statistics are interesting in that they show that the large majority of clients (74 per cent) are experienced in construction and are therefore likely to have a significant effect on how their projects are organised.

Their conclusions are that the conventional wisdom of a large number of 'naive' clients of the construction industry, who really do not understand the nature of the process and its inherent difficulties, is misplaced. They believe that the industry is one in which there are a sizeable number of regular clients whose average project is one in which they have considerable experience. Such clients typically manage a fair-sized portfolio of projects varying in size and type, and will often have some in-house capacity and well-established mechanisms and procedures for handling them.

Masterman and Gameson (1994) draw on the work of Higgins and Jessop (1965), and Nahapiet and Nahapiet (1985) in their classification of clients which they base upon:

(1) whether clients are primary or secondary constructors
(2) clients' level of construction experience.

Their definitions of constructors are:

Primary: 'Clients such as property developers, whose main business and primary income derive from constructing buildings.'
Secondary: 'Clients for whom expenditure on constructing buildings is a small percentage of their total turnover, and for whom buildings are necessary in order to undertake a specific business activity, such as manufacturing.'

Levels of construction experience are defined as:

Experienced: 'Recent and relevant experience of constructing certain types of buildings, with established access to construction expertise either in-house or externally.'
Inexperienced: 'No recent and relevant experience of constructing buildings, with no established access to construction expertise.'

When these two characteristics are considered together the following four alternative client types are produced:

- Primary experienced
- Primary inexperienced
- Secondary experienced
- Secondary inexperienced.

Whilst such a framework is useful it should not be allowed to disguise the fact that the classification identifies four basic types amongst which there exists a large variety of gradations which can be represented as in Fig. 5.1.

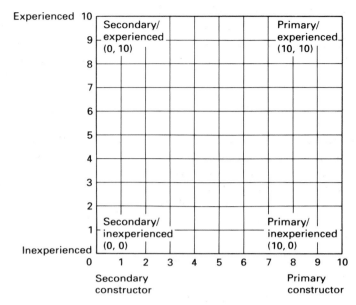

Fig. 5.1 Graduation of client types (developed from Masterman & Gameson 1994).

Clients' objectives

The most important feature of any building project should be the client's objectives in embarking on the construction of the project. The need for the project will normally have arisen from some demand arising from the client organisation's primary activity. For example, the client's primary activity may be food processing. In order to remain competitive the client may need to absorb within the firm work previously subcontracted, which may require construction work to be undertaken. A client with a plant located overseas may wish to provide it with its own power supply in the event of an unstable political situation threatening existing sources. The client may be an education authority and there may be a demand for additional school places, which have to be provided by the construction of a new school. A health authority may be required to respond to advances in medical science by providing a new treatment facility.

The needs of clients will therefore be stimulated by the environment of their organisation, which will present opportunities to which they respond. Such external stimuli may be economic forces, which give the opportunity for profit, or sociological forces, which present the chance to respond to a social need, but more usually they are combinations of different classes of stimulus. The basic response of a client to environment forces is the result of the need to survive; above this level the client responds in order to expand as a result of drive and motivation. Survival is the basic objective of clients and can be defined as maintaining their position relative to those of their competitors. This is more easily conceived for commercial organisations but it is also true for public clients. In commercial terms it requires sufficient response to stay in business. In terms of non-profit organisations, it means sufficient response to prevent the organisation being replaced by some other mechanism: for example, the establishment of urban development corporations to undertake some of the work of certain local authorities. A further example is the establishment of housing associations to provide some public housing previously provided by local authorities.

The effect of forces in the client's environment will therefore trigger the start of the construction process although it may not be realised at the time that a project is needed and at that stage it is unlikely that any member of the project team will be involved. When it becomes apparent that a construction project is needed to satisfy the client's objectives, the brief begins to be formed. A common major problem is that the project team will normally not be involved at this early stage and a number of important decisions which may inappropriately constrain the design of the project may have been made by the time they are brought in.

The client's brief is often perceived as a reasonably detailed statement of what the client requires but it is important that the strategic level of the brief is not overlooked at the expense of detail. The client will be concerned in the three major areas of price, quality and time. The weight different clients give to each of these factors will vary. They will require value for money but on their own particular terms. A public client with a low capital budget but high revenue budget may wish to suppress the initial price at the expense of quality; a client building a prestigious building may prefer the opposite; a client in a rich, developing country may see the time for completion as paramount at the expense of price but not of quality, and so on.

Quality and price can be subdivided into components, each of which will have its own weighting within the balance which a particular client wishes to achieve between competing factors. A summary of a typical weighting is illustrated in Fig. 5.2. The balance which the client requires may not be possible and compromise between conflicting factors may have to be negotiated. For instance, it would be illogical to have a low weighting for technical standards and a high weighting for low life cycle costs. Ward *et al.* (1991) draw attention to the trade-offs which are necessary between these interdependent factors. They believe that the problem of trade-offs is in most cases complicated by uncertainty about the nature of the interdependencies between the different value-for-money criteria. They point out that the pair-wise effects for the three basic criteria of time, cost and quality do not always work in one direction and may depend on circumstances. For example a decrease in project duration may lead to an increase in cost but it does not have to, it may cause a decrease.

This type of strategic scenario will be required as a backcloth against which the detailed brief can be prepared. The client's priorities will therefore have to be established. It may well be that there is conflict within the client's organisation regarding priorities. The project team will need to be confident that it has

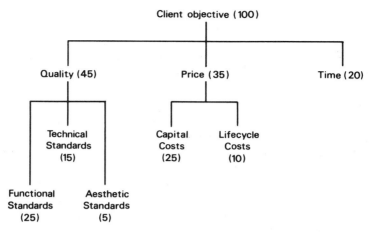

Fig. 5.2 Client objectives – weighting of factors.

interpreted the balance properly and to achieve this it will have to understand the client's organisation, its decision-making process and where its highest authority lies. It will be against such a concept that clients will ultimately judge their satisfaction with the completed project and upon which the reputation of the project team will rest. This is not to say that it is easy to make clients understand the conflicting pressures of a construction project. Most clients would expect each component to be weighted at 100 per cent. Given unlimited capital this may be possible, but is not the reality of most construction projects.

In a comprehensive review of work on client objectives Walker, D.H.T. (1994) summarises the lessons learned as:

'● Clients with a detailed and firm idea on what needs to be done and how this can be achieved in a well planned and appropriately controlled manner can be viewed as sophisticated and will probably assist in a successful project outcome.

● Clients who are novices or unsure of how to brief principal consultants may be successful if they know the limits of their expertise in this matter and know how and when to ask for appropriate assistance.

● Clients must be firm in the statement and direction of their objectives and they must speak with one voice. This need has led to the rise of the role of the project manager.

● Clients should remain involved in the whole design and construction process to ensure that work is being undertaken effectively and that they can contribute positively throughout the process.'

These are the lessons learned by the investigators. The trick is how to have the lessons learned by clients and project team members. As a corollary Bresnen and Haslam (1991) found that there is a tendency for at least some of the problems of additional work (over and above the original scope of the project) to be the result of inadequacies in the original brief.

Relationship of the client's organisation and the construction process

An organisation can be considered as an open adaptive system in terms of a general input – output model as in Fig. 5.3. Both the client's organisation and the construction project organisation can be considered in this manner.

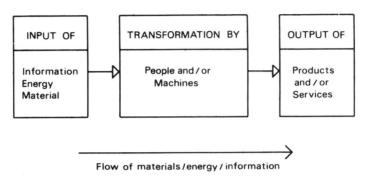

Fig. 5.3   General model of an organisation as an open system.

An open system is in continual interaction with its environment and retains the capacity for work or energy transformation. A system must receive sufficient input of resources from its environment to maintain its operations and also to export the transformed resources to the environment in sufficient quantity to continue the cycle. For example, the client's organisation, whether private or public, receives inputs from society (its environment) in the form of people, materials, money, information, etc. It transforms these into outputs of products, services and rewards to the organisational members that are sufficiently large to maintain their participation. The output is therefore returned to society (the environment) in some form. The project organisation performs in the same way although the nature of the inputs and outputs differs.

In Chapter 1 it was suggested that the management system of an organisation could be seen to consist of:

- the organisation sub-system
- the behavioural sub-system
- the technical sub-system
- the decision-making sub-system.

The technical sub-system is defined by the technology required to undertake the task of the organisation and is represented by the skills, knowledge and equipment required and the way in which they have to be used. Although it can be developed and adapted by the organisation, it is frequently prescribed by the current external state of development of the particular process. The technical sub-system is that to which the behavioural sub-system has to relate and with which it must be integrated. The behavioural sub-system will have a significant influence on the effectiveness of the utilisation of technology. The organisation sub-system is the structure that relates the technical to the behavioural sub-systems, and the decision-making sub-system is the mechanism through which the managing system activates the organisation.

The implication of this scenario is that analysis of the technical system will produce a systematic picture of the task and tasks relationships of an organisation to which the other sub-systems relate. Such a view of a client's organisation allows the project team to understand the *modus operandi* of its client's organisation and gives it a basis for integration during the construction process. The technical sub-system of a client can be readily perceived in manufacturing industry (e.g. manufacturing vehicles, electrical components, producing oil, etc.) but it is equally applicable to non-manufacturing activities, for example treating patients (health authority), collecting taxes (inland revenue), designing advertisements (advertising agency), etc.

This scenario also applies to the construction process itself. The technical sub-system is the technology required for designing and constructing the project. The behavioural sub-system is the attitudes and values of the members of the process. The organisation sub-system is the way in which they relate to each other and the decision-making sub-system is the mechanism through which the process moves forward. The technical sub-system dominates and in this respect it is important to recognise the differences in the technical sub-systems between construction projects. For example, the technical sub-system for the construction of a house is quite different from that for a multistorey car park, a theatre or a power station, and so on. It therefore follows that the organisation and decision-making sub-systems should be designed to reflect the technical and behavioural sub-systems. This demands that a variety of organisation solutions should be available to suit the particular project. There are therefore two systems involved, that of the client and that of the construction process, and they become joined temporarily for the duration of the project. The construction process becomes a temporary sub-system of the client's organisation, as shown in Fig. 5.4.

The client's primary activity can be seen as an input – transformation – output system and a response to environmental forces triggers the start point of the construction process. A part of the input to the client's primary activity (e.g. money, energy) is diverted to become an input to the construction process, which will also acquire other inputs directly from its environment. In both cases the inputs can be summarised as materials, information and energy. (Energy is the input that drives the transformation process and therefore includes people, ideas, power, etc.) The output of the construction process will then return to the transformation process of the client's system to provide an additional facility, which will contribute to the primary activity and assist in achieving the client's

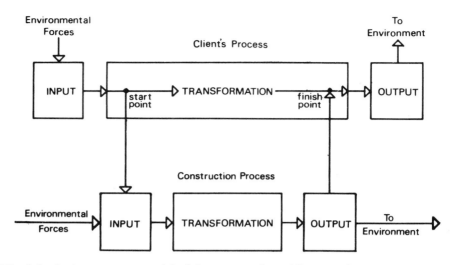

Fig. 5.4 An input–output model of the process of providing a project.

objectives. The construction process can therefore be conceived as an internal transformation within the client's system and as a temporary sub-system of it.

Conflicting objectives

It has been widely recognised that business organisation systems tend to have multiple objectives and that some form of compromise takes place. Multiple objectives arise as a result of the network of relationships that exist within a system. This is particularly to be expected for construction projects owing to the client–construction process relationship and the fact that the construction process itself often consists of a number of organisationally independent firms. The benefit of taking a systems view is that the conflicting multiple objective situation can be made explicit.

Multiple objectives arise because of the individual aspirations of the sub-systems (e.g. firms or departments), which tend to develop their own purpose outside the main purpose of the system. It is therefore important to identify and relate a system to predominant objectives. In terms of the client–construction process relationship, the predominant objective must be that of the client, which will reflect the primary function of his organisation.

One of the tasks of the management process is to ensure that sub-systems remain orientated to the primary function of the system. A company must be sure that its sub-systems (e.g. departments) are not developing discrete objectives that conflict with the company's primary objective. A good example of this is what is often referred to as 'empire building', in which managers of departments are concerned only with building up their own department irrespective of optimising its contribution to the firm. The possibility of conflicting objectives in the construction process is even more likely as not only may the client's sub-systems develop discrete objectives and try to implement them during the briefing of the project team, but the sub-systems (e.g. firms) making up the project team may also develop discrete objectives that conflict with the client's objectives. An important role of project management will be to ensure that the objectives of the client are accurately and clearly stated and that all the contributors to the project remain orientated towards them.

The primary function of the client's organisation will be that process which it continuously undertakes in order to survive as an organisation. The construction process has its own primary function which is providing the project, which should remain compatible with the primary function of the client. Circumstances may arise in which the primary function of the construction process temporarily becomes the primary function of the client's system but normally it will remain subservient to it. For example, if it is known that the only way by which the client's firm can survive is by completing and commissioning a building so that a particular process can come on stream by a particular date, the client's primary function will, temporarily, be to ensure that this happens. A danger following a temporary shift of the primary function is that it may lead to a permanent redefinition to the detriment of an organisation's ability to survive. Similarly, if the leaders or members of an enterprise do not agree on their definition of the primary function, the survival of the enterprise will be at risk.

Orientation of the construction process towards its objective of providing what the client requires is achieved through feedback. The client's objective for the project and the details of how this is to be achieved should be stated in the brief. Feedback loops should be designed into the construction process to establish whether the output of the process is compatible with the brief. Such feedback points should coincide with the major decision points in the process and be designed to ensure that any additional information arising from the environment during design and construction that may require the brief to be amended is taken into account.

Project change

Most construction projects are designed and constructed in accordance with the brief established early in the process, but the project team needs to be aware that uncertainty and change in the client's environment may require that alterations have to be made to the project in order to respond to them. The Royal Institute of British Architects (1991) Plan of Work indicates a time after which the project brief should not be modified. This is rather idealistic as it is surely not reasonable or sustainable to try and tell a client who is investing a considerable amount of money that the design cannot be changed when the operating circumstances have altered significantly. However even the RIBA Plan of Work recognises that in reality changes will be made but points out that changes of the detail design will result in abortive work.

An understanding of the state of the client's environment is necessary for the project team, since, in conditions of uncertainty of the client's needs, flexibility must be maintained. This may be achieved, for example, by not selecting a long-term fixed solution such as a new building, or by designing a building that exhibits the flexibility demanded by the environment. The implication of this is that the maximum amount of up-to-date information regarding the requirements of the client must be maintained when the project is being designed. For example, an advance in technology for a particular process (e.g. high-rise fork-lift truck design) could mean that significant amendments to the brief should be made to take advantage of such developments.

Conversely, changes in the environment of the construction process should be allowed to amend the structure of the process if this can be done to the advantage of the client. The members of the project team should keep themselves aware of any such changes and be ready to advise their client accordingly. For example, shortage of steel could result in a change of design or in ordering the steel prior to signing of the main building contract, or in bringing forward the commencement of construction by adopting innovative approaches to appointing a contractor.

Role of the client

The degree of involvement and the role of the client in the construction process will depend to a large extent upon:

* the structure of the client's organisation
* the client's knowledge and experience of the construction process
* the authority vested in the various levels of the client's organisation
* the personal characteristics of the client's people who have responsibility for the project.

If the client constructs frequently, there will probably be employees familiar with the process who can act as co-ordinators on the client's behalf and who will liaise within the organisation and between it and the project team. In such a case the client will maintain a presence close to the project. The effectiveness of this arrangement will depend on the degree of authority vested in such a co-ordinator. If this is high, it will be to the advantage of the project provided that the co-ordinator has personal skills and characteristics that gain the confidence of the project team. In these circumstances the delegation of authority to the manager of the project team by the client will not need to be high. On the other hand, if the co-ordinator does not possess much authority and/or has personal characteristics that are not suitable, this arrangement may be counterproductive as it could lead to frustration and delay.

The authority of the co-ordinator will depend upon the authority pattern within the client's organisation. If the organisation is mechanistic, it is probable that authority will not be delegated to any great degree. The project team will therefore have to rely on the higher levels of the client's organisation for decisions. This could result in delay in reaching decisions, as it is unlikely that people at such a level in the client's organisation would have sufficient time in addition to their other activities to devote to keeping close to the project. If, however, the client's organisation is structured so that authority is delegated well down the organisation, a member of the client's staff who is intimately involved with the particular project may have authority over most matters. This should result in close integration of the client organisation and the construction process, with the effect that decisions can be readily obtained to the advantage of that process.

Much has been written about the amount of authority which the 'project manager' should or should not be given. In the context of a project manager who is external to the client's organisation, this is a difficult decision for the client. The project team will be spending a large amount of the client's money and human nature does not normally allow some external agency the authority to do this. Nevertheless, in such complex circumstances as a construction project, such decisions have to be faced and resolved. Much will depend upon the structure of the client's organisation and its experience of construction. As referred to earlier, if the client appoints its representative from within its own organisation and gives that person a large degree of authority, then the manager of the project team should not require much authority as there will be easy access to the client's representative. On the other hand, if the client keeps authority for the project high within the organisation, vesting significant authority in the manager of the project team should be considered. The problems start to arise when neither of these situations exists.

Similarly, problems will arise when authority is not clearly defined as often happens when the client fudges both authority and responsibility for the project within its organisation. This usually 'knocks on' to an unclear statement of the authority of the manager of the project team. Problems then manifest themselves as either decisions not taken at the right time or even not at all, or in decisions being frequently changed leading to delay and abortive work.

There is no set solution to the integration of the client and the construction process. Each mechanism will have to be designed to suit the particular organisation of the client body. The client cannot be expected to change the organisation structure fundamentally because of a temporary embarking on construction. Therefore the mechanism should reflect the organisation, yet clearly state the pattern of authority and responsibility for the project. Only from this basis can the decision-making process and communication system for the project be identified. The integration of the client and the project team is a most significant factor in the success of the project and requires understanding and skill in its design.

Kometa *et al.* (1994) show that certain attributes of clients other than client integration contribute to a successful project performance through the manner in which clients' attributes influence the performance of consultants. The attributes identified include financial stability, feasibility, feasibility of the project, past performance of the client, project characteristics and client duties (which did not include integration). This is of course a different perspective, concerned with the substance of the client body rather than management, but is nevertheless interesting. Complementing these findings were those of Walker, D.H.T. (1994) which discovered that project team confidence in the client and the client's representative rather than vice versa reflects the difficult role a client's representative has to play as a link-pin between a multidimensional client group with conflicting goals and the project team.

Clients and projects

Project teams tend to start developing projects assuming that the client has:

(a) identified the best means of achieving its objectives
(b) carefully analysed the spatial, technical and performance requirements associated with its objectives.

The data provided by the client is therefore frequently accepted without question as the basis for developing the design. The inappropriateness of such assumptions is illustrated by the redevelopment of a ferry pier in Hong Kong. Redevelopment was considered appropriate as the surrounding area had been redeveloped including reclamation of land which was subsequently developed with residential units. As a result a new transport complex was built, into which the ferry pier was fully integrated. It was decided that some commercial facilities should be made available to attract passengers. The resultant design, which was quite novel and the first of its kind in Hong Kong, comprised the ferry pier portion that could accommodate simultaneously two triple-deck ferries and two hovercraft. The pier incorporated two further levels, one for a shopping arcade and public promenade and the other for a restaurant. The 'assumption', without detailed study, was that the demand for ferry service would increase in view of the proposed residential developments on the new reclamation. Unfortunately, because of a new subway line, the actual demand for ferry service since the ferry pier was commissioned was far from that assumed. As a result there has been insufficient patronage of the commercial development and arguments resulted over lease conditions. The commercial areas proved not viable and were eventually converted to offices. The world is littered with similar examples of lack of objectivity and no matter how efficiently resources are applied in devising and executing designs and construction, if they are not achieving realistic objectives the inevitable result is waste.

Whilst lack of an objective evaluation will invariably lead to unrealistic objectives, the internal politics of the client organisation can contribute equally to a lack of objectivity, distortion of objectives and potential problems for the project team. This process was well illustrated by Cherns and Bryant (1984):

'Consider the case of the corporate client that always has more projects competing for finance than it has capital available. Thus the agreement to invest in X (say, building a new warehouse) is a decision which not only pre-empts X^1 (extending an existing one) but also gives priority over Y (building a new laboratory) or Z (installing a computerised production system). Since Y, Z and X^1 are projects in competition with X, and all have their organisational supporters, the decision for X is a victory for the supporters of X – and a defeat for the supporters of the other competing projects. (Examples of client interest groups supporting competing projects include operating divisions, specialist departments, political factions, professional groups, etc.) The victory for X then marks a shift in the balance of power within the client organisation (if only to confirm the dominance of one interest group over another). But the victor has now given a hostage to fortune. In fighting the battles, he has probably shaded the risks and been optimistic in his estimate of cost in order to get his budget just below the organisation's threshold of approval for capital expenditure. He has promised a return and now has to deliver – or lose credibility.

The study has provided us with vivid glimpses of the various "hostages to fortune" that the different interest groups within the client complex may have offered in promoting a particular project in competition with others. Each participant can be seen as bringing to the table his own sense of what is at risk personally, as well as what is at stake professionally or departmentally, in the forthcoming project experience. Some client participants have a high stake in meeting the target completion date; others in working within the promised budgetary limit; others in the operating performance of cherished design concepts. Many of the stakes are reputational (e.g. the operational manager's reputation as a skilled negotiator in the best interests of his own department; the project manager's reputation as a hard-headed realist who gets results in the face of whatever obstacles).

In considering the role of the client in construction, then, we cannot treat the client as unitary; nor we can ignore the events which preceded the decision to build. The progress of a construction project involves various groups within the client organisation whose interests differ and may be in conflict, and whose observed behaviour cannot adequately be explained without reference to the past.'

It is important that project teams not only find the client's objectives realistic, but also that they have some understanding of the organisational dynamics which brought them forward. In order to be reassured on the former and have knowledge of the latter there needs to be a high level of trust and compatibility between the client and project team leader. The team leader will need to ask the client many searching questions before the brief is fully developed. It is debatable whether many clients will either be prepared to or be in a position to satisfactorily answer such questions. As a result many client objectives are unsatisfactory and lead to unsatisfactory projects for which the project teams are likely to carry a large part of the blame if not the responsibility. There is a great need for a high level of understanding of this process on the part of both clients and project teams.

Cherns and Bryant's study identified a number of tentative propositions about client involvement in the construction process of which the following are particularly pertinent to the points developed above:

- Most client systems are very much more complex organisationally (in terms of who wants the building, who will use it, who approves it, who controls the money, etc.) than is commonly acknowledged by project teams.
- Members of project teams seem to be impatient of this complexity, sometimes even embarrassed by it, and insist on dealing with a single client representative within whom all the internal politics of the client system can be contained.
- Many of the problems concerning design changes, delays, and difficulties during the construction phase have their origins in the unresolved conflicts within the client organisation and are exacerbated by too early an insistence on an over-simplified client representative function.
- The earliest decisions taken by the client system have more influence over the way the project organisation is formed and its subsequent performance than those taken later.
- The reasons for these earliest decisions have their origins in the client's organisational culture, procedures and structures; they are often idiosyncratic, shaped by social and political forces as well as by economics and technical considerations, and may be unduly constrained by residues of the client's pre-project history.
- Any serious attempt to understand and hence to disseminate usable knowledge about the client role in the organisation and management of construction project must take into account not only technical and economic factors but also the social and political forces acting within the client organisation, especially the influence of the client's pre-project history on the decision to build.
- The decision to build is a large-scale innovative decision with consequences for existing patterns of resource sharing and risk taking in terms of power conflicts and political behaviour within the client organisation.
- These conflicts and behaviours can critically affect the formation, development and subsequent performance of the project organisation which is set up to manage the project and of which the client system is an initiating component.

These propositions clearly illustrate the issues facing project teams. Whilst in most cases the problems they generate are unlikely to be resolved, their importance lies in project teams knowing of their existence and being prepared to understand and adapt to the dynamics at work to the benefit of the project outcome.

One particular project analysed by Walker and Hughes (1987b) is illustrative of the organisational complexity of client organisations. The project was a divisional police headquarters constructed on a confined town centre site. Technically, the client was a Metropolitan County Council. In reality the client was multi-headed and, in addition to the Council, consisted of the Home Office, the Police Committee, the Chief Constable and representatives of the police force. Compounding the difficulties of dealing with a complex client body was lack of sound integration between the client body and the project team. The complexity of the client body is illustrated by the fact that 40 *different* management units, many of which arose from the various branches of the client organisation, had, at some stage of the project, the responsibility to approve some aspect of the project. There were also 30 different units with the power to make recommendations. The completed project was unsatisfactory due in large part to the complex nature of the client which generated procedures which imposed a 'dead hand' on the project.

Allen (1984) illustrates the benefit of the project team adopting a more positive and creative approach to defining the client's needs. He does not commence from the more conventional standpoint of both client and project team which he believes is:

'• The client has already decided that he needs a building and has already identified his site.
• The client draws up a schedule of his requirements for that building, both in spatial and performance terms. Sometimes, a layout of his spatial requirements is also provided.
• The design team produces (maybe alternative) outline proposals, in response to the client's requirements, complete with an indication of specification, programme and cost. This may, or may not, meet with the client's approval. If it does, the project proceeds through its remaining stages. If it doesn't meet with approval, it may have to be modified, to varying degrees, or started all over again.'

This process assumes that the client knows precisely what is required and all the project team needs to do is work out the details and arrange for the project to be constructed. Allen believes that it is normally wrong to make such an assumption. He illustrates the benefits of a more dynamic approach to clients' needs using the following projects:

'(i)   The client approached his project team with the intention of extending an existing laboratory complex in two places; continuing its tendency to 'grow like Topsy'. Upon investigation, it became apparent that the client had given little consideration to his future needs; both in terms of expansion and his day-to-day operations. The result of the project team's initiative was an overall development plan, involving a rolling series of building phases. Some of these included refurbishment of the existing buildings, to accommodate changing occupational patterns as the development plan unfolded. In operational terms, the entire laboratory complex became much more effective and efficient. Inevitably, this led to relatively greater profits for the client – which, after all, was his original objective.

Had the design team adopted the more traditional approach to this particular client's initial brief, the two extensions proposed would have only served to further compound the relative ineffectiveness and inefficiency of the existing complex.

(ii) A major vehicle manufacturer approached his project team with the intention of building a new engineering centre for the whole of his car design activities. The client's initial intentions were to build a separate office block, workshop block, styling centre, etc., to house his existing departments. Adopting the approach advocated, the design team sought answers to the question 'How are (or should) motor cars (be) designed?' The building design, which emerged from this process, was substantially a single building, with each of the functions involved in designing a motor car sensibly related. The other consequence of the process was that the entire organisational structure of the company's engineering centre would change, to reflect the clarity which the briefing process had produced.

Again, this is an example of the design team seizing the initiative, in the interest of satisfying the client's real objective. In this instance, the effect of that initiative extended far beyond their immediate function as a design team.

(iii) By way of contrast, a Polytechnic, which, in spite of stringent cash limits, sought and achieved a similar level of service to the public as was achieved elsewhere, where cash was not so limited. The design team's approach was to give careful consideration to the use of space. By careful timetabling and the resultant greater use of each space, the total amount of new space required was significantly reduced. This, inevitably, reduced the capital investment required, without loss of service to the community.

This serves as an example of where the client had a very clear objective and had the wisdom to involve the project team in identifying the most economic means by which that objective could be achieved.'

The key to success, in each example, was the initiative and perception of the particular project team or, in the latter case, the client. In all three examples, had a more traditional approach been adopted, it is extremely unlikely that the client's needs would have been met by the resultant designs.

Peter Morris (1994) describes many of the world's 'mega-projects' in his historical account of project management including Atlas, Polaris, F-111, Apollo, Concorde, the Trans Alaska Pipeline and others. The role of the client (or 'owner' which he prefers) is strikingly portrayed in his accounts in which project definition plays a crucial role for either good or ill, particularly in cases when the client is that most complex of 'owners' – government.

These examples illustrate the complexity of the relationships between clients and project teams in which many millions of pounds are frequently involved. A successful project inevitably means that both the client and the project team leader have to work in a spirit of trust, openness, collaboration and creativity to identify the appropriate objectives for the project and so give it the greatest chance of success.

Chapter 6
The Project Team

The firms

Reference was made in Chapter 1 to the large number of firms involved in the construction process and in Chapter 3 to the theoretical basis of their formation. These firms are often independent units which are interdependent in terms of the work they undertake – the design and production of construction work. Even so they vary considerably in the range and quality of skills they offer. The number of different combinations of firms that may be involved in a construction project is extremely large. As a result, firms have to be familiar with working with a variety of other firms, and at any one time within a particular firm many different amalgams of firms will be working on the projects in hand.

Table 6.1 gives some examples of the types of amalgam that may exist, and the additional variety introduced by the various ways by which the construction contract may be awarded. The table illustrates eighteen primary examples of different arrangements, but many more variations are available. Each of the different arrangements generates a different set of relationships between the contributors. At any given time a person can find themselves involved in a variety of situations as it is not unusual for someone to be working on a number of projects simultaneously. The complexity of the situation is compounded by the variety of clients and projects which overlie the professional relationships.

Within the amalgams shown in Table 6.1, management takes place at various levels. In each case the individual contributing firms have to be managed. The partners or directors will be concerned to maximise the efficiency of their firm while at the same time enhancing its reputation for service. The same will also apply to departments of public authorities. If firms are concerned with more than one skill, e.g. a professional consortium and local authority architects' departments, this will involve managing not only the individual skills but also the collective skills of the members. Within individual firms, the service provided to a particular project will have to be managed within the context of the total firm. Resources will have to be allocated to satisfy the demands of the project and also be within the capacity of the firm, and decisions regarding both the quality and quantity of resources allocated must be made. For example, the services engineer will have to manage the services engineering provisions within the resources of

Table 6.1 Examples of amalgams of firms.

Designers / Contractor	Appointed by competition after design is substantially completed	Appointed by negotiation or competition early in design process	Management contract	'In-house' to client	Design-and-build
General practice surveyor, architect, quantity surveyor, structural engineer, service engineer, all in separate private practices	X	X	X		
As above, plus a project manager in separate practice	X	X	X		
Consortium of design skills including project manager and separate general practice surveyor firm	X	X	X		
All design skills 'in-house' to client (e.g. local authority, private developer), including project manager	X	X	X	X	
Some design skills including project manager, 'in-house' to client with others in separate private practices	X	X	X	X	
'In-house' to contractor					X*

* In addition the client may appoint consultants to oversee the contract on its behalf

the services engineering practice, the contractor the construction process, etc. At a lower level each individual contributor will have to manage its own contribution. All these activities will have an objective in terms of satisfying the client, but they will also have other objectives. The dominant one will be to ensure a profit for their firm (this also applies to public authorities in terms of effectiveness) while at the same time maintaining the firm's reputation. Individuals will have personal objectives (e.g. self advancement, avoidance of responsibility). While the objectives of the firm, the project and the individuals continue to be satisfied simultaneously, all will be well. However, if these objectives clash, it is the role of the manager of the project on behalf of the client to resolve the conflict in the client's interest.

Project management is the management of the contributors to the project who will be from different firms except where they are totally 'in-house' to the client's organisation. Its sole objective is the satisfactory completion of the project on behalf of the client. It therefore normally crosses firms' boundaries and for its purposes temporary management structures are created for the duration of the project. They will be disbanded on completion of the project but may be re-formed for future projects. It is important that the contributing firms recognise the existence of temporary management structures and organise themselves so that their members become full members of those structures. This requires that firms be prepared to dedicate staff to projects even though this may at times appear to be to the detriment of the internal organisation of the firm. Firms should therefore be structured to allow staff to give allegiance to the projects on which they are employed and to be involved with only the number of projects that allows them to devote sufficient time to each. Dedication to projects should enhance a firm's reputation, and dedication to firms should enhance profit in the short run. If both are not achievable, a trade-off between them takes place to the detriment of one or both. Such a situation demands sensitive and skilful management if both reputation and profit are to be optimised and if staff are to be motivated and retain job satisfaction. Staff should be seen to be rewarded for achieving the appropriate balance.

If this balance is not achieved, the project team has been described (Association of Project Managers 1984) as having 'a limited objective and lifespan, and therefore with a built-in death wish'. The same publication also saw it as potentially a 'weak system compared with the continuous and self-perpetuating drives of the other systems. Putting together effective teams in such circumstances from a wide variety of organisations and motives is therefore a difficult and complex task'.

The general management literature refers to the structure which emerges as a matrix and its management as matrix management. Mainstream management writers tend to find a matrix difficult to relate to, with comments such as 'much ambiguity must be tolerated and competing claims accommodated for the matrix to function. For many participants, matrix structures are high demand, high stress work environments' (Scott 1992). Yet, construction professionals, contractors, subcontractors and clients have always existed in such structures as they represent the basic structure which is inevitable as a result of the task to be performed.

Criticisms of the matrix structures are cited in Poirot's (1991) paper as over complex, difficulties of quality-control and operational difficulties particularly when applied in large companies. But herein lies the difference as much of the mainstream management literature focuses on matrix management within a single organisation rather than that required by construction project management which relates to inter-organisational management. Nevertheless Poirot continues by showing that in his own multidisciplinary consulting firm the matrix structure has been most effective and that:

'Almost any organisation structure will work if the users accept it, but with the matrix system, managers must have clear definitions of authority and responsibility. They must have mutual respect for each other and must be excellent communicators. They must be able to set aside their individualism and understand that the essence of a matrix is that the group must succeed first, then the individual.

The matrix system is a power-sharing, power-balancing organisation. If all the power moves to one side of the matrix, the whole organisation loses. Most decisions are made at lower levels of management in order to push decision making as far down as possible in the organisation and to encourage participative decisions.

Our matrix works! We feel the decisions creating our matrix have resulted in continuous growth and the ability to weather recessions without major decreases in our permanent staff.'

Further insights to the differences between inter-organisational and intra-organisational matrix management are provided by Robins (1993). He drew attention to the fact that in-house project managers always have the authority to control subcontract expenditure (which can also be interpreted as consultants in construction) when work packages are subcontracted out but not when the work is undertaken by in-house functional departments. In the latter case he believes that separation of responsibility and authority occurs and that effective management control is prevented. He believes that if a functional department is seen as a subcontracting business within the organisation and accounting procedures are set up to recognise this, then the problem will be solved.

It becomes the job of the in-house functional managers (subcontractors) to sell their services to the project manager against competition from external subcontractors (or consultants). These views highlight the difficulties of in-house matrix management as opposed to matrix management of external organisations. Although the latter overcome many of the criticisms in the general management literature they can of course bring problems of their own such as contract disputes and adversarial attitudes.

Relationship with the client

The complexity of project management structures raises the issue of how best the project team should be integrated with the client's organisation. The two ends of the spectrum are illustrated by, at one end, the project manager or other leader of the project team as the only point of contact between the team and the client, with all instructions and advice being passed through this channel. At the other end, all members of the project team have direct access to the client and in this arrangement the leader of the project team co-ordinates the instructions and advice given. Either of these alternatives is likely to be unsatisfactory in most cases and the appropriate integrating mechanism will probably lie somewhere between.

The design of the mechanism will depend to a large degree on the amount of authority delegated by the client to the manager of the project team. Where substantial authority is delegated, most of the contact between the client and the project team is likely to be directly with the manager of the project team acting as the surrogate client. In such cases, for aspects for which the manager of the team does not have authority to act, the client may well prefer the leader of the project team alone to present the recommendations. In cases where the manager has little authority, the client may prefer to thrash out issues in consultation with all or some of the members of the project team. Alternatively, the client may require the project team manager to make recommendations for a decision, so that the client can discuss them with the whole group. In this arrangement, although the manager may not have delegated authority from the client, significant authority is gained from being in the position of co-ordinating and influencing the pro-position(s) that form the basis of recommendations to the client. In putting prepositions and recommendations to the client, the project team manager will also be in a good position, even in the presence of other members of the team, to influence the final outcome.

Where there is direct access by team members to the client, the team manager will need to ensure that the client receives a balanced view and that decisions are made in the light of all the factors affecting the project, rather than as a result of the statements of the strongest personality in the team. For example, many construction projects do not have a formal, detailed brief. The detailed brief emerges through the architect placing a series of sketch designs before the client, which are amended or rejected. The 'brief' therefore proceeds incrementally until the client 'sees what is required' on the drawings. In such a process it is essential that the other members of the team be present and involved. Otherwise important elements, such as cost and time constraints and certain significant elements of design (e.g. aspects of services), may be ignored through con-centrating attention upon other aspects. The result could be that a building is proposed which the client cannot afford and which, by that stage of design, cannot be amended in time to be completed to programme. Such a development is potentially more likely if the architect or other professional contributor is also the project team manager than if the leader were independent of the professional contributors.

There can be no hard and fast rules laid down for integration with the client. So much will depend upon the particular views held by the client and the client's experience of construction projects. However, the project team manager can influence most clients and should ensure that whatever is devised is clearly laid down and understood by everyone involved, particularly the client. The essence of integration is that the decisions made as a result of contact with the client are controlled in terms of the objectives of the project. Unilateral decisions made by either the client or one of the team can, at best, lead to confusion which will take a great deal of unravelling and cause abortive work. At worst they may be incor-porated into the project, with the result that whereas they may satisfy one aspect of the project's objectives, they defeat one or more of the client's other objectives, which in the long run may be more significant to the client's satisfaction with the total project.

Differentiation, interdependency and integration

The specialisation of the contributors to construction projects has been occurring throughout the world since the last century. As observed in Chapter 2, in the UK from the basis of architect/builder have evolved quantity surveyors, various specialist engineers, general contractors, specialist subcontractors and general practice surveyors. Even within these specialist occupations there are often further specialist subdivisions. For instance, there are design architects, detailing architects and job architects; in the quantity surveying field there are building economists, bill preparers and final account specialists. Whereas on some projects the same person may undertake all the functions of a particular contributor, for many projects a number of people are involved at the different stages of a particular specialist contribution. Add to this the way in which clients' organisations are often subdivided into specialist groups, all of which have a contribution to make in terms of project definition, and the complexity of the interrelationships that emerge is clearly evident. In systems terminology such specialisation is called differentiation (Lawrence & Lorsch 1967), which can be defined in construction terms as the difference in cognitive and emotional orientation among contributors to projects who offer specialist skills. The differences in cognitive and emotional orientation of the specialists within the construction process are readily apparent. Each of the specialists tends to view his or her colleagues with a certain amount of scorn. The contractor rarely expresses respect for the architect and vice versa, and no one has a good word for the quantity surveyor or consulting engineer! The divisions are wide but narrowing, nevertheless, at project level there exists a pressing need to ensure that such differences are reconciled so that they do not affect the performance of the project team to the detriment of the project and hence the client.

Closely related to the concept of differentiation is the concept of sentience, developed by the Tavistock group (Miller & Rice 1967). A sentient group is one to which individuals are prepared to commit themselves and on which they depend for emotional support. In the construction process this can arise from allegiance to a firm or to a profession or to both. It is a particularly strong force in the construction process and it is from sentience that the perception of the process by the different contributors arises. It has been found that sentience is likely to be strongest where the boundaries of a sentient group and of a task coincide. This is the usual situation in the construction process. For instance, traditionally, architects have normally been solely involved in architecture and builders in building, with very little, if any, overlap. The various contributors have a tendency to focus upon and be concerned only with their own specialism and are unable to perceive and respond to the problems of others.

Sentience is weakest in a group of unskilled or semi-skilled workers whose roles are interchangeable and where each individual is dispensable. Individuals in such a group will not acquire sentience unless the group finds supplementary activities through which members can make individual and complementary contributions. It has been found that sentience is strongest in members of a professional body that confers upon its members the right to engage in professional relations with clients in which task and sentient boundaries coincide. There is a specific danger when both direct relations with clients and coincidence of boundaries of sentient and task groups occur in that it may produce a group that becomes committed to a particular way of doing things. Although both efficiency and satisfaction may be greater in the short run, in the long run such a group is likely to inhibit change and behave as though its objective had become the defence of an obsolescent method of working. This view appears to have some significance for the construction process. There have been many pleas over the years for the boundaries between the professions of the building industry to be broken down but there is still little evidence of this having taken place on any appreciable scale.

A phenomenon which has similarities with sentience is the concept of groupthink (Janis 1983) which is a process which can work against effective group decision making. Whilst conformity in groups (which encompasses project teams) which are charged with decision making is necessary to enable progress to be made, excessive conformity interferes with constructive critical analysis and leads to ineffective decision making. Weak arguments and uncritical thinking go unchallenged in the group in order not to disturb the stability of the group. The members of groups which turn out this way subject themselves to social pressures, self-censorship, illusions of unvulnerability and unanimity, rationalisation of decisions and have self-appointed mind guards. Project managers need to guard against groupthink in their project teams. Methods of counteracting groupthink are available (Leigh 1983), one of which is 'teamthink'. Teamthink is directed at self-managing teams and compares with groupthink as follows (Manz & Neck 1995):

Groupthink

Description: Group members striving to agree with one another, overwhelms adequate discussion of alternative courses of action. Defective decision-making results.

Symptoms:
- Direct or social pressure against divergent views
- Self-censorship of concerns
- Illusion of invulnerability to failure
- Illusion of unanimity
- Self-appointed mind guards that screen out external information
- Collective efforts to rationalise
- Stereotyped views of enemy leaders
- Illusion of morality.

Teamthink

Description: Groups engage in effective synergistic thinking through the effective management of its internal dialogue, mental imagery, and beliefs and assumptions. Enhanced decision-making and team performance result.

Symptoms:
- Encouragement of divergent views
- Open expression of concerns/ideas
- Awareness of limitations/threats
- Recognition of members' uniqueness
- Recognition of views outside the group
- Discussion of collective doubts
- Utilisation of non-stereotypical views
- Recognition of ethical and moral consequences of decisions.

The differentiation of skills together with their reinforcing sentience, can be clearly seen in the construction industry. It is also quite clear that all the contributors, each in their own 'box', are interdependent in carrying out their work of producing the completed project to the client's satisfaction. The network of interdependencies is practically total. It is not that each contributor is dependent on one other but that all contributors are in some way dependent upon all the others. If members of the process were asked if they were interdependent, they would undoubtedly agree, but this is not something that would be at the forefront of their minds if the question were not asked. This lack of recognition of interdependency begins with the education of members of the construction process. Each discipline is educated in relative isolation from the others. Exceptionally there may be some joint work, but if so it is only likely to be a very small proportion of the time devoted to study. The difficulties of resolving this situation are compounded by different patterns of courses and a lack of will to break the mould and reform the educational process. This problem is particularly apparent in the UK and was referred to once again in the latest government report (Latham 1994), but is common throughout the world.

The contributors are interdependent because on the one hand the various tasks that have to be undertaken to achieve the finished project require an input from a range of contributors, and on the other hand because the tasks themselves are interdependent as frequently a task cannot be commenced until another has been completed or unless another task is undertaken in parallel. Different types of interdependency exist and have been classified as pooled, sequential and reciprocal (Thompson 1967).

Pooled interdependency is basic to any organisation. Each part renders a discrete contribution to the whole. The parts do not have to be operationally dependent upon or even interact with other parts, but the failure of any one part can threaten the whole and therefore the other parts, for example the decentralised divisions of a large, diversified company. In the construction process, if one part fails, it will not necessarily mean the failure of the other parts but the failure may reflect upon the reputation of the other parts. *Sequential interdependency* takes a serial form. Direct interdependency between the parts can be identified and the order of the interdependency can be specified. For example, bills of quantities must be prepared before tenders can be invited (using this particular form of procedure). *Reciprocal interdependency* is when the outputs of each part become the inputs for the others and the process moves forwards through a series of steps. Each step requires interaction between the parts and each part is penetrated by the others. This is seen, for example, when preparing an outline proposal for a building which must be functionally and aesthetically sound and also feasible from a structural and cost point of view.

The three types of interdependency can be arranged in ascending order of complexity as pooled, sequential and reciprocal. A more complex type also contains the less complex types. The order of complexity is also the order of most difficulty of integration. Integration (Lawrence & Lorsch 1967) can be defined in construction terms as the quality of the state of collaboration that exists among contributors to projects who offer specialist skills and who are required to achieve unity of effort by the environment. If, therefore, there are different types of interdependency, there would need to be different methods of integration. As reciprocal interdependency is the most difficult to integrate, and as this type of interdependency dominates in the construction process, the integrative mechanisms and effort need to be of a high order.

It has been found that the integration of pooled interdependency is best achieved through standardisation and formal rules, and sequential inter-dependency through planning. That is, the tasks to be undertaken can be anticipated and their sequence planned so that sequential interdependency is identified and recognised at an early stage. The managing process should then ensure that integration takes place as planned. Reciprocal interdependency is integrated by mutual adjustment and feedback between the parts involved.

The integration of reciprocal interdependency requires close association between interdependent parts to ensure that the required input takes place at the appropriate time and that account is taken of the various inputs in the process. The management of reciprocal interdependencies requires that a balance be maintained between the inputs in terms of clients' objectives. A clear perception of clients' objectives is required, together with the diplomacy and expertise necessary to integrate a group of highly skilled professionals.

A study by Mohsini and Davidson (1992) attempted to measure the impact of some organisational variables on the performance of projects which used the traditional building process. They found that two factors, the sufficiency of starting information which stresses the importance of information exchange and the extent of the tasks' interdependence, were consistently important. Inter-dependency is, of course, generated by the differentiation needed to cope with task and environmental complexity. Whilst recognising that structures without a central integrating mechanism were not suitable for large and complex building projects, they also suggested that newer procurement strategies should seek to reduce the negative impact of 'task interdependence'. They suggest that the use of process-related innovations which reduce the number of participants would be of benefit, for example design-and-build and reducing on-site operations using prefabrication. However, whilst these are useful suggestions, the opportunity to reduce interdependency are limited as it is determined predominantly by the nature of the task and the impact of the environment in the project, both of which are tending to become more complex.

In an organisation which is strongly differentiated yet largely interdependent, such as that found for construction projects, the key to success is the quality and extent of the integrative effort provided by the managers of the organisation. The root of project management should therefore be the integration of the organi-sation. This applies whatever the organisation structure adopted. Within any organisation there will be someone, or some group, responsible for managing the process. Conventionally, the architect both designed and managed. Increasingly, a project manager is appointed to manage the process. In either of these cases, or any other, the manager's fundamental activity is integration.

The mainstream literature on organisation still finds the use of task forces and project teams as 'complex and fascinating' (Scott 1992). The predominant concern is often their relationship with the hierarchy, the high differentiation within specialist project teams and the need for strong integrative efforts which are seen as exceptional arrangements. These views are also reflected in general management's disquiet with matrix management structures. This is opposite to construction project management for which such structures are *de rigueur* and hence second nature to professionals in the industry. For these reasons the general management literature requires reinterpreting to distil that which is useful for project team organisation and management.

Integration will be necessary in two specific directions. One is the integration of the people involved with each specific task. At a basic level the manager will need to ensure that the appropriate people with the right skills are involved at the correct time. This may seem obvious but it is surprising how often this does not happen. If advice is given too late in the process, or if it is not given at all or not taken, it can lead to abortive work or delay and dissatisfaction by the client with the outcome of the project. So within each task the integration of the contributors needs to be ensured. This type of integration is achieved at a personal level through the characteristics of the manager. The manager should create in the project team recognition and respect for the contribution of others by all members so that the member responsible for carrying out a specific task automatically seeks advice. The manager should then monitor progress within tasks to ensure that the development of the project reflects the level of response between contributors considered to be necessary.

The other, and equally important, need for integration is between the output of the tasks. Each task undertaken by the project team will have to be compatible with each other and in relation to the project's objectives. The manager of the process, who should be taking an overview of the various tasks being undertaken, should evaluate the output of the tasks in terms of their compatibility. This will require what is in effect a feedback loop at each output at which the latter is 'measured' against the project objectives and against the output of other tasks. For instance, it is unacceptable for the design of the electrical services to satisfy the client's objectives for power and lighting if the proposed solution cannot be provided within the cost limit for the work. Adjustment would have to be made to the design, the cost limit, or the client's objectives. Sound integration within tasks should be designed to prevent such an occurrence, but integration of the output (between tasks) acts as a 'backstop' with formal feedback opportunities to ensure that within-task integration has taken place effectively.

The integrative mechanisms designed into the organisation structure will depend on the particular project and its environment, but will range from integration through personality to formal and rigorous feedback mechanisms at key points. This is understandable as differentiation is present within the system for two main reasons. One of these is to do with the emotional orientation or characteristics of the people involved in the process and the other is related to the technical nature of the projects themselves, which are often complex and demand input from a number of skills to tasks that have to be combined to produce the completed project.

Decisions and their effect on structure

As referred to in the last chapter, the decision-making process and specifically the timing of decisions made by the client can have a significant influence on the effectiveness of the project organisation. The decisions the client makes will be based upon recommendations or alternatives presented by the project team, usually through its manager. Such decision points act as 'pinch points' through which the project must pass if it is to progress. If an acceptable decision cannot be made, the project will not squeeze through the 'pinch points' and will terminate, or the objectives will be revised. Between these major decision points will be others at which decisions will not normally be made by the client but by the project team manager, depending upon the authority pattern that has been established.

Decisions taken at the project team level will contribute to those taken by the client and at each level the manager of the project team needs to integrate the various tasks to produce the alternative propositions available. The manager will then make the decision when it is within his or her authority or make a recommendation to the client if the latter is to make the decision. In order to stand the best chance of making the correct decision, the range of available alternatives and supporting arguments will have to be presented in each case. It is the manager's role to make sure that all the alternatives are exposed and to achieve this he or she will have to take account of the advice of all the contributors to the project. Integration of the contributors therefore assumes paramount importance. The integration of contributors within and between tasks is important, but the key integrating activity of the manager is in bringing together the output of the tasks in a way that allows the range of available alternatives to be clearly exposed.

Thus the essential determinant of the structure of an organisation for the design and construction of a project is the arrangement of decision points and the way in which the contributors need to be integrated in order to produce the material upon which decisions can be made. The most significant decisions are taken by the client and the timing and sequence of the decision points will be determined by the internal organisation and external environment of the client's organisation. The framework needs to be elicited by the manager of the project team from the client before the positions of the decision points for decisions within his authority can be determined. The manager will then be in a position to design the integration of the contributors in the project organisation for the purpose of both classes of decision. Whereas it may be possible to identify a list of routine decisions that are common to all construction projects, it is not possible to determine when they will need to be taken until the framework of the client's decision points is known. Nor is it possible to identify non-routine decisions within the authority of the manager until the client's decision framework is established. Given knowledge of that and of its level of certainty, managers are able to integrate decisions and the contribution of their team in the most effective manner.

A major task of the manager of the project will therefore be to make the client realise the fundamental nature of its role in the construction process and the way in which it can affect the effectiveness of the process and the client's own satisfaction with the project outcome.

Differentiation and integration in practice

Differentiation and attempts to integrate are expressed in a whole variety of ways in practice. Experience on one project with a particular group of participants is not necessarily transferable, either to another project or to another group of participants. Each project should be analysed individually to identify the type and scale of differentiation as a basis for designing the appropriate organisation structure and integrating mechanism. Later chapters will examine how this may be achieved but at this point it may be useful to look at some of the problems and solutions which have been commonly tried in practice.

The most positive approach has been the creation of multidisciplinary practices that employ within the one firm all the professional skills associated with projects. If, within such practices, specialists work in project-dedicated teams, then one would expect that conditions would be created in which a high level of integration could occur. However, if such practices continue to organise in 'departments' of specialist skills, a great integrating opportunity will have been lost. In either case, integration with the contractor will be difficult to achieve if the contractor is not appointed until after design has been substantially completed. Even if appointed early in the process, differentiation will be high and special integrating effort will be required. A similar situation will exist on a development project if, as often occurs, a general practice (valuation) surveyor is involved from outside the consortium.

Differentiation is high on a project when professional consultants are from separate firms and they will be differentiated from the contractor to varying degrees depending upon when and how they are appointed. If positive attempts are not made to integrate them, the effect upon the project outcome can be serious. A recent example encountered was a project for which the quantity surveyor and services engineer never met but communicated by post and telephone. The result was that the services installation cost control was badly managed, leading to abortive work and dissatisfaction by the client. A more positive approach was discovered on another project on which the professional consultants, although from separate practices, worked together for the whole of the design phase in the office of the consulting engineer who also provided the project management service. At first the consultants did not like the idea of uprooting themselves and working in unfamiliar surroundings but after the event they agreed that it had been very beneficial in creating a harmonious team and producing a project that met the client's requirements.

The familiarity of the contributors with each other's methods as a result of working together on previous projects does, of course, assist integration but can lead to complacency. This can be evident where the same team works on a subsequent project, which places different demands upon the team as a result of the client's requirements and the environmental influences acting upon the process. There is a danger that there may be no stimulant within the team for a change of approach.

As mentioned previously, the problem of integrating the contractor is always present. Many of the new initiatives of contractor involvement in the design team are intended to try to assist the integration of the contractor into the project team. Design-and-build, management contracting and the USA construction management approach are examples of initiatives that bring the designing and constructing functions closer together. Although the benefit claimed for these approaches is that they allow an input of construction knowledge to the design, there is a potential for an equally important benefit in terms of integrating the people working on the project.

Design-and-build has emerged over recent years as an increasingly popular procurement method. Its appeal to clients arises from single point responsibility which simplifies the manner in which the client interacts with the project team. Theoretically a design-and-build organisation structure should reduce differentiation and provide a sound platform for effective integration resulting in a proficient management structure. However, in practice this may only be the case if all the project skills are in-house to the design-and-build company. If this is the case differentiation will be reduced and an in-house project management culture can be developed leading to high integration. But frequently design-and-build companies do not have all professional skills in-house. There are a number of reasons for this, including the desire to limit investment and the problem of retaining the scope of skills (particularly design) which may be required. As a result professional skills are hired in from individual professional practices. The design-and-build company in such circumstances, which will be based on contracting, will have the responsibility for integrating the project team. The organisational advantages in such circumstances is questionable although the client still has the advantage of single point responsibility. The design-and-build company will need highly developed project management skills as there may be a reluctance on the part of members of professional practices to be managed by construction companies unless the practices are carefully selected. The client may also retain a project manager and other professional advisors to oversee the design-and-build company. If professional skills are in-house to the design-and-build company the relationship with the client's advisors should not be too difficult to manage but if the design-and-build company hires in professional advisors the situation becomes much more complex due to the multiplicity of firms involved.

The appointment of a project manager on behalf of the client, either in-house to the client organisation or from an independent firm, should act basically as an integrating device although the benefit of such an appointment is often justified in other terms, such as progress chaser, controller, etc. or even just as a pre-occupation with titles. One would expect that the greater the differentiation between the contributors, the greater the need for a project manager. The latent differentiation of contributors to all projects as a result of sentience and other forces means that the need for integrating effort is always high and would probably benefit from the integrative effort provided by someone solely concerned with project management.

Most building projects require someone to act as a catalyst. This need is often recognised for large complex projects and there is no doubt that the scale and complexity of such projects, both technically and environmentally, expose differentiation and demand an integrating mechanism such as a project manager. However, the extent of differentiation on the medium sized or even the smaller projects is not as readily recognised but is sufficient to require positive action to integrate the contributors rather than just hoping that it will happen.

International projects can generate the greatest levels of differentiation. Not only will the contributors be differentiated for the reasons given previously, but such differentiation will be compounded by the contributors coming from a variety of countries and being required to apply their skills in a country with which they may not be very familiar. The differences in cultural background and methods of working will generate differentiation, which can only be integrated by careful organisation design and very positive effort. It is therefore not surprising that project managers are employed more frequently on international projects than on national ones. This need is also reinforced on overseas projects by the environmental and often technical complexity of the work, as projects are frequently undertaken to build whole industries and extensive facilities from scratch in conditions of great uncertainty.

Theoretically, to reduce differentiation to a minimum, clients would develop their projects using a team of specialist skills as employees within their own organisation ('in-house') including the construction phase using directly employed labour. In such a situation the likelihood of conflicting objectives among the contributors would be reduced. The allegiance of the contributors should be largely to the client directly, although allegiance to professional skills would not be eliminated altogether. In this type of arrangement the opportunity to generate the maximum level of integration using an in-house project manager, who would have access to and a full understanding of the organisation's objectives, should be at its greatest. However, in most cases this is just not practicable as most clients do not have such an in-house capability. Local authority direct labour organisations in the UK were one of the few examples of this type of arrangement but their history is not generally one of great success, although there are some successful examples. Perhaps their problem was more political than management orientated. Nevertheless, the in-house scenario is an interesting one upon which to base thinking about the type and degree of differentiation and integration present in any specific project organisation.

'Partnering' is a recent major initiative aimed at greater integration of the project team. Its focus is behavioural rather than structural as it aims to change the traditionally adversarial relationships particularly between clients and contractors but also encompassing the whole project team. The objective is to achieve the project by working together constructively rather than by confrontation. Partnering is defined by the US Construction Industry Institute (1989) as:

> 'A long-term commitment between two or more organisations for the purpose of achieving specific business objectives by maximising the effectiveness of each participant's resources. This requires changing traditional relationships to a shared culture without regard to organisational boundaries. The relationship is based on trust, dedication to common goals, and understanding each other's individual expectations and values. Expected benefits include improved efficiency and cost effectiveness, increased opportunity for innovation, and continuous improvement of quality products and services.'

To make partnering work requires great commitment from all parties and a belief that a new approach to relationships is in everyone's interests. To achieve this requires education of the parties on partnering, commitment of top management, the appointment of a 'champion' by each party to the agreement to implement the concept within their organisation, a partnering workshop which develops formal agreements and processes for the resolution of issues and disputes.

It is important to recognise that partnering does not eliminate the need to structure the project organisation effectively but there is no doubt that if this is achieved partnering can be of enormous benefit to the manager of the project through the way in which it influences the behaviour of the participants and eases, or overcomes, problems of integration and elimination, or reduction, of disputes through the key themes of teamwork, collaboration, trust, openness and mutual respect (Larson 1995). Partnering could be seen as a condition precedent to sound integration of project teams.

Larson studied 280 construction projects in the USA to identify the relationship between project success and alternative approaches to managing the owner–contractor relationship, of which one was partnering. The results showed that partnered projects achieved superior results in controlling costs, technical performance and in satisfying customers compared with those managed in what was termed adversarial, guarded adversarial and on an informal partnered basis. Whilst he points out that the study was exploratory and has limitations, nevertheless the results are important. He predicted the inferior results for adversarial and guarded adversarial approaches but found the differences between informal and fully-pledged partnering significant and well worth the investment in full partnering.

Management techniques and project information

Recent years have seen the development of a range of techniques aimed at making the management of construction projects more effective and at providing clients with the means to make more rational and objective decisions. This has been accompanied by the rapid growth of cheap computing power useful to such techniques. However, there is often a gulf between the availability of the array of tools for project control and their actual use on real projects. The opportunity and will to apply techniques depend upon a receptive management structure and an appropriate configuration of contributors to a project. Ability to organise to take advantage of the growth in techniques and computing power has fallen behind the rate of development of technology. This is particularly so for construction projects for which organisational structures still tend to reflect the juxtaposition of traditional professional roles. This inhibits innovation with the result that the industry and professions are slower than many other industries in applying new ideas and techniques. Techniques such as networking, life cycle cost planning or even cash flow forecasting are still often instigated and implemented in response to the clients rather than by the professions and industry themselves.

Such a lack of initiative often occurs due to the fact that there is no person who is solely concerned with the overall management of the construction process. Each contributor is predominantly concerned with their own particular part of the process and is therefore unlikely to generate the use of techniques that have implications for the project as a whole. For example, traditionally, the architect had overall responsibility for management as well as for design and monitoring of construction, but the architect's allegiance to architecture tended to make the architect place overall project responsibilities at a lower priority than architectural matters. In addition, architectural education did not generally equip the architect with the necessary perspectives for such a role.

A person charged with the sole task of project control is likely to recognise a need for project control techniques and to evaluate and implement those techniques that will help them in the task of overall management of a project. An organisation structure that provides this mechanism should itself initiate the implementation of innovative techniques rather than solely responding to client pressure. A shift in perception from a focus on the parts of the construction process carried out by the specialist contributors to an overview of the process as a whole will not only improve service to clients and the effectiveness of the process but will also raise the reputation and status of the industry and its professions.

The techniques and skills used in the construction process rely fundamentally upon the nature and quality of the information they use and the manner in which the information is structured to make it appropriate and accessible to each particular activity of the project organisation. The co-ordination of project information has been the subject of much research attention in the UK culminating in the publication of the Co-ordinated Project Information (CPI 1987) documents. So often information produced during the construction process has related to the specific needs of the information generator rather than being useful to the process as a whole. Hence the architect produces drawings and specifications which are themselves often unco-ordinated; the engineers do the same and their drawings are often not co-ordinated with those of the architect. The quantity surveyor translates this information into cost plans and bills of quantities, which are again often not co-ordinated between themselves or with the drawings from which they were derived. The contractor uses the bills to tender, and then has to translate the information given in the bills and on the drawings, etc. into a form suitable for use in construction. None of these data are usually co-ordinated with programming and cash flow information.

One of the most important recommendations of the Latham Report (Latham 1994) was that the use of Co-ordinated Project Information should be a compulsory part of the conditions of engagement of a designer. The report remarks that co-ordination of documents seems basic common sense and should have become normal practice years ago. One client commented. 'If Knowledge Based Engineering is tomorrow's technology for construction, CPI ought to be yesterday's. Surely we can harmonise the basic works information?' The report also believes that CPI is both practicable and desirable when design information is not complete before construction commences.

Recognition that not all information required for construction projects is of a technical nature is important. Information input to decisions in the early development of a project will be at a much more macro level and be concerned with defining the client's requirements. This process will need information generated by the client's company and its environment and information about the scope of design possibilities, the capability of the construction industry to deliver what is required, and the procurement paths available.

The responsibility for designing and implementing an information system for a project rests with the person responsible for managing that project for the client. A more positive approach to project management and project organisation design could lead to a greater use of management techniques and to the design and adoption of project information systems that contribute to effective, co-ordinated communication between participants and provide the data to make management techniques more viable and helpful.

Chapter 7
A Model of the Construction Process

Introduction

It will be useful at this point to draw together the threads that have been running through the previous chapters into a model of the construction process. This approach formalises the ideas that underpin the way in which construction project organisations should be structured and provides an approach to analysing and designing project management structures. Although the approach may at first appear theoretical, it does provide the basis of a practical, analytical tool for examining the effectiveness of the project management process, as described in Chapter 12.

The construction process has few fundamental characteristics common to all projects. This is not unexpected in view of the diversity of construction projects and their clients. That being the case, it is necessary to identify those aspects that can be generalised so that they may then be interpreted for each individual project. The application of the model will then identify the structure of the process in such a way that it is possible to analyse how it operates in practice.

Such a model may be employed as a tool for learning from experience in a more rigorous way than has been the case in the past, by using it to analyse completed projects. But, more importantly, it can also be used for designing organisation structures with the aim of providing the structure which should give the best chance of a successful project outcome as far as the organisational aspects are concerned. The tasks and roles of project management can then be identified on the basis of the organisational structure designed for each specific project. Project success is, of course, dependent upon much more than solely organisational issues, such as behavioural, political and other forces acting upon the project, but if the organisation structure is as well designed as possible, at least the project is off to a good start.

Common characteristics

A prerequisite of the model is an outline of the process of providing a project devoid of artificial organisation boundaries such as those created by conventional and other predetermined approaches to project organisation. Such a model would identify the major forces that influence the process and the fundamental structure that results.

The process has a start point (which may be difficult to identify specifically in practice). It also has a finish point, which is taken as the completion of a project. The process of identifying and providing a project consists of those events that join these two points. Potential start points are activated by organisations which *may* become clients of the construction industry if the process identifies that a construction project is required to meet the objectives of the potential client. The term 'client' is used to refer to a sponsor of construction work who can generate the finance, information and authority necessary to embark upon the process.

Construction projects start as a result of the influence of environmental stimuli upon prospective client organisations which create the motivation and need or opportunity to construct to reach objectives. Such stimuli may be economic, technological, sociological, etc. and usually consist of combinations of different classes of forces. The basic response of an organisation to environmental stimuli is the result of its need to survive; above this level the organisation responds in order to expand as the result of its motivation (see Fig. 7.1). Survival is the basic goal which requires the organisation to maintain its position relative to those of its competitors for which it must continue to obtain a return acceptable to its environment in terms of its role (e.g. profit, service, acceptability). This is more easily conceived for commercial organisations, but is also true for public authorities.

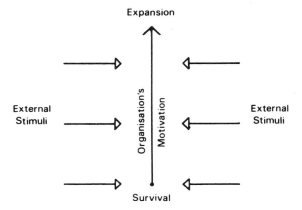

Fig. 7.1   Organisation's response to external influences.

Public authorities have to adapt to survive and also strive to expand by increasing the quality of their services. If they are perceived not to be doing so other means of providing their services may be found. Privatisation of previously public services in many western economies is evidence of this although many such changes have not been as a result of economic forces as proposed by the trans-action cost approach but rather as a result of political pressures.

Expansion is a response to environmental forces by the organisation to take advantage of events in its environment. The degree to which the organisation takes such opportunities is determined by its motivation, which is, in turn, influenced by incentives provided by the environment, e.g. taxation, status, satisfaction. The start point of the process is, therefore, the recognition by the potential client of the need or opportunity to achieve a particular objective for their organisation. The options available to achieve the objective *may* include the acquisition of real property (defined here as encompassing both building and civil engineering outputs) which, in turn, *may* require the construction of a new project, but at this early stage this will not have been established.

At the initial activation of a start point, the plane within which a finish point is feasible will be very wide and will encompass all those alternatives that allow the organisation to achieve its objective. The alternatives available will vary, depending upon the nature of the organisation's role. For instance, there will be basic differences between the choices available to commercial and public authorities. However, *it is possible*, for every category, that one of the alternatives will require the acquisition of real property.

This outline of the process is now illustrated further using as an example the identification and provision of a building for a commercial organisation. The concepts are more readily understood in terms of commercial criteria, although the same process should be followed by any potential client and for processes which lead to either a building or civil engineering project as the outcome.

After starting the process, the initial decision of relevance to the construction process is whether or not real property is required. This may be called the *project conception process*, as illustrated in Fig. 7.2. If a decision is reached that does not require the construction of a project, then the organisation which was a potential client of the construction industry will not become a client. During this phase, environmental influences are transmitted to the potential client through the importation of information, energy and materials from the environment. The meaning of information and material is self-evident, although it should be pointed out that material encompasses any material whatsoever. Energy similarly means any type of energy, but in this context people are a particularly important source of both physical and mental energy. Such influences can be broadly classified as political, legal, economic, institutional, sociological and technical. The action of these influences will determine the initial decision. The project conception process will entail the consideration of each alternative within the environmental context and a decision will be made on the basis of the influence of the external factors. For example, economic conditions may make a process change appropriate, but it may then be discovered that trade union action (sociological influence) will make this difficult, by which time economic conditions may have made the take-over of another firm more appropriate. This process is one of the client organisation adapting to environmental influences

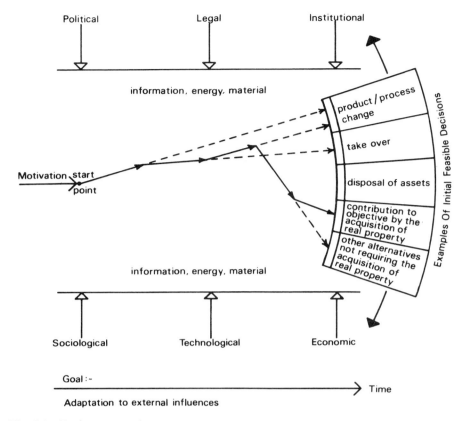

Fig. 7.2 Project conception process.

until an initial feasible decision is reached and normally takes place within the client organisation.

In developing the model, it is assumed that the preferred outcome of the project conception stage requires the acquisition of real property to contribute to the satisfaction of the potential client's objective. At this stage, acquisition of real property includes existing or new property or improvement or modification of property already owned.

The assumed preferred decision of the project conception process, which requires the acquisition of real property, contains a number of alternatives which can be considered as an intermediate feasible decision point. The process of arriving at one of these alternatives in making further progress towards the finish point may be called the *project inception process*, and is illustrated in Fig. 7.3. The intermediate feasible decision actually made is again determined by the ability of the alternative chosen to contribute to the achievement of the objective of what is now a client of the construction process. The environmental influences acting upon the process of reaching an intermediate decision are the same as those given

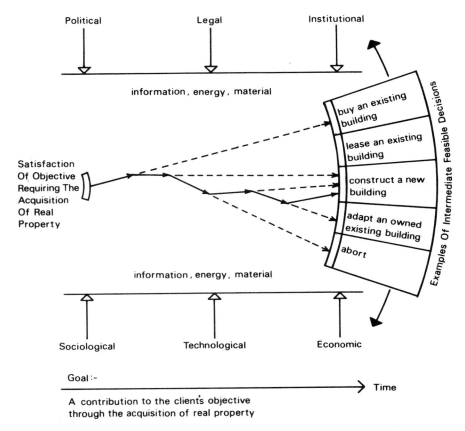

Fig. 7.3 Project inception process.

before, but may exert different influences during this process. The project inception process will receive information, energy and material from the environment and will transform them in its task of identifying the appropriate intermediate decision. Interacting with these influences in arriving at a decision will be the commercial activity of the client, which will itself be influenced by the external factors. For example, the environment factors affecting the decision will be the state of the property market regarding the availability of existing premises and rent levels, the cost of new building work, site availability, the rate of technological change which determines whether a short-term lease is better than a freehold building, etc.

In developing the model further, it is assumed that the preferred outcome of the project inception process is the construction of a new building. The performance of the building that is actually constructed will lie within a finish spectrum ranging from total satisfaction with performance requirement to total dissatisfaction. The process of arriving at the finished building is called the *project realisation process*, as illustrated in Fig. 7.4. This will again be determined by the

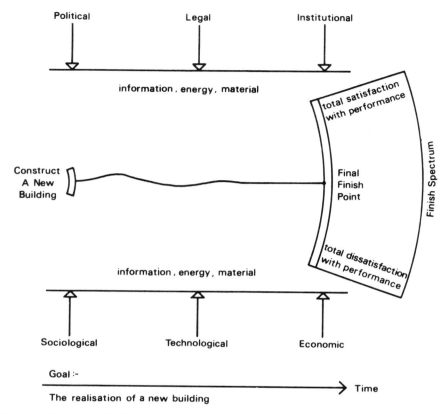

Fig. 7.4   Project realisation process.

environmental influences acting on the process. These are classified as before, and again provide information, energy and material for the process. For example, the environment provides the skills, both professional and constructional, which are available to the project. It also determines the availability of materials, and even the weather, which might affect the completion time for the building. The client's environment will affect the certainty of what is required of the building. If it creates uncertainty, this might generate changes to the design and construction which can affect the cost and completion time for the project. As was the case with the project conception and inception processes, the external influences act in two ways: directly upon the process and indirectly through their influence upon the commercial activity of the client. The project realisation process transforms these inputs into the output of the process, which is the finished building. The effectiveness of the transformation process will determine the quality of the outcome actually achieved.

A further example of the effect of external influences during this process could be that economic and/or institutional forces determine that construction work is awarded on the basis of a competitive tender. Such a decision would divide this process into two sequential sub-processes, design and construction, but only if appropriate external influences are present. Such an assumption would be unfounded at this stage of development of the outline of the process.

To summarise, there are only three sub-systems that have universal application to all construction projects: the project conception, project inception and project realisation sub-systems. These sub-systems generate two primary decision points, one which contains the potential decision that real property is required and, if such a decision is taken, the second division as to the nature of the real property to be provided. Both these decisions will be taken by the client organisation.

The start point represents the beginning of the project management process and it can be seen that this will, in most cases, be contained and managed within the client organisation during the project conception sub-system. Ideally, the client organisation should, during this process, receive advice from the project team, but in reality this rarely happens. If it were to occur, then the members of the construction team involved would be part of the project management process but would not be leading it, as at this stage a business decision rather than a construction decision is required. However, it would certainly be advantageous for clients to have an input of advice from certain members of the construction team at this stage to enable them to take a fully informed decision.

The project inception sub-system demands a significant input from the project team and the process will require a property or construction orientated management system working in conjunction with the client to identify the most appropriate solution to the client's needs. However, project management does not usually take place in this form in practice. It tends to occur implicitly, solely within the client organisation, and project teams are often faced with a *fait accompli* by the client. This may be appropriate when the client has in-house expertise but to the disadvantage of the client when this is not the case. Project teams can, therefore, often do little other than proceed on the basis of the client's preconceived idea of the best solution to the problem. It would be advantageous to both the client, in terms of the utility of the completed building, and the project team, through their increased effectiveness, if clients were to involve members of the project team in this process.

The project realisation sub-system is the process that is most readily perceived by clients and project teams as that in which project management takes place, although even in this process it is sometimes construed as being concerned only with the construction phase rather than with the whole process, including design. Even though project management may be led by someone from the project team during this process, the management of the process will benefit significantly from the involvement of the client.

As there are only three sub-systems that are universally applicable to construction projects, it is necessary to identify the factors that determine the sub-systems within each of these major sub-systems which are identified by the two primary decisions discussed previously. As most projects differ, the sub-systems required to achieve them will also differ as a result of the task being undertaken and the environment within which it is carried out. Therefore further sub-systems cannot be defined explicitly, but what is possible is to identify the factors that create them, the nature of their relationships and their need for integration.

Sub-systems

The primary decision points differentiate the major sub-systems and define the boundaries between them. Similarly, other decision points will determine the boundaries between other sub-systems. The construction process is characterised by a series of decision points. These act as 'pinch points' through which the project team must pass if progress is to be made. If the project cannot pass a pinch point, then it will be aborted. Decision points are arranged in a hierarchy of which the primary decision points are at the highest level. Below them occur key decision points and operational decision points, the distinction between which Bennett (1991) has recognised that their identification and the distinction between them as being especially important in providing the basis for integration.

Key decision points

Further discontinuity in the system is created by key decision points. Key decisions are those that the clients will make for themselves. They are determined by environmental influences acting upon the client organisation and are often manifest in the client's internal procedures for expenditure and similar approvals. They can range from, for example, approval of design and budget proposals and decisions to delay the project, to decisions to change the nature of the project. Such decisions imply a degree of irrevocability, as to revoke such decisions would entail the client in a loss of resources. Bennett (1991) confirms that experienced clients often ask for a formal report at key decision points.

The process of providing a project is characterised by discontinuity created by the need for decisions. As they reflect the flow of the process they are fundamental to the organisation structure of the project. The integrating mechanism provided between the client and the members of the project team is, therefore, highly significant for the success of the project. This mechanism should aim to anticipate key decision points and organise so that decisions can be made on the basis of the contribution of the members of the project team, who have a part to play in providing information and advice on which such decisions should be based. The nature of the client's organisation can have a fundamental influence on the effectiveness of this process. If the client is represented by a committee or board of directors, they will have to decide how they themselves are to take key decisions or whether they will delegate this function to some member of their own organisation and, if so, to what extent they will delegate. Similarly, they will have to decide upon the extent of delegated authority given to the project team as this will determine which are the key decisions. Decisions taken by the project team are classified as *operational decisions*.

The greater the number of decisions classified as operational, the greater will be the flexibility available to the project team and the more control they will have over the decision-making process. There is likely to be less uncertainty and delay and more integration, for example, in those cases where a client's representative has sufficient delegated authority to make many key decisions, or in cases where the project team deals directly with, say, a managing director acting with full authority. This is likely to be the situation in the former case only where the client has great confidence in the representative and, in the latter case, for the smaller private company. Very real difficulties can be created for the larger private companies and public authorities if an appropriate integrating mechanism is not or cannot be designed. A common problem is that of key decisions taking longer than anticipated, with a resultant delay to the project, which can often have a corresponding knock-on effect for later project activities. A further common problem is that if the client's organisation is unresponsive to environmental forces and the needs of the construction process, key decision points may be inappropriately identified in terms of the nature of the decision and its timing. An important task of the leader of the project team is, therefore, to endeavour to make clear to the client the team's needs in terms of the timing and quality of key decisions. Yet, at the same time, the leader must recognise that the project will place heavy demands upon the client's organisation, which will have to continue to carry out its main function as well as being involved in the design and construction of the project.

Key decisions lie beneath the primary decisions in the decision hierarchy and, therefore, contribute to them. Until a primary decision point is reached and the decision is taken, key decisions will be cumulative. They therefore provide major feedback opportunities for both the client and the project team. Each key decision should be checked against the overall objective of the project and the primary and key decisions already taken, to ensure that the project is remaining on course. If it is not, then it will either have to be returned to course or the original objectives or previous decisions will have to be reassessed to establish whether the deviation is beneficial and feasible. Whereas the primary decisions create the boundaries between the major systems of the process, the key decisions represent the boundaries between the main sub-systems which constitute the major systems and provide feedback opportunities as shown in Fig. 7.5.

Operational decision points

In bringing forward propositions upon which key decisions will be based, the members of the project team will themselves have to make decisions based upon their professional and technical competence. These will not affect the policy of the client's organisation: such decisions will be primary and key decisions, and will be taken by the client. Therefore the decisions taken by the project team in

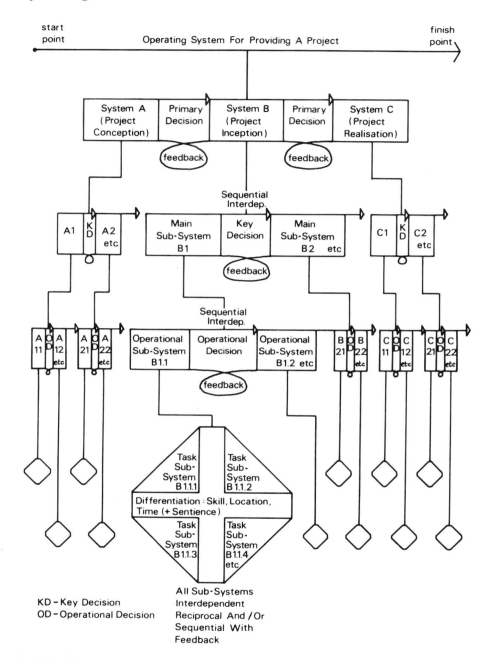

Fig. 7.5 The operating system.

making progress towards a key decision are described as operational. The range of propositions that may be presented to a client, and the operational decisions implicit in defining the propositions, will be the responsibility of the leader of the project team. The leader will also have the responsibility for integrating the project team to ensure that all relevant advice has been given before arriving at an operational decision. The leader must also make certain that the full range of propositions appropriate to the client's objectives have been formulated and that they are presented objectively at key decision points. Unfortunately in practice, on many projects clients are not presented with alternative propositions and in such cases operational decisions are made solely to move the process towards the next key decision point at which the client is asked for a 'yes' or 'no' decision to the single proposition.

As with key decision points, operational decision points will also represent 'pinch points' through which the project must pass if progress is to be made. Although they do not have the same degree of irrevocability and associated loss of resources as key decisions, nevertheless operational decisions which are later changed are likely to cause delay and some loss of resources. Examples of operational decisions include the details of project programmes, the use of bills of quantities for tender documentation, and alternative technical design proposals. These decisions are therefore mainly concerned with professional decisions and with the implementation of procedural aspects of projects, and move the project incrementally towards a key decision. Their position in the system will be determined by previously taken key decisions, but there will be more opportunity for the project team to design the structure of these decision points than is the case with key decision points.

Operational decision points present secondary feedback opportunities. Each time an operational decision is being considered, it should be checked against the previous operational decisions within the sub-system and against the last key decision to ensure that it is compatible with what has been decided previously. If it is incompatible, opportunity to change the decision exists or, alternatively, to amend a previous operational or key decision, although this is unlikely to be the outcome at an operational decision point.

Each sub-system created by key decisions will, therefore, consist of a number of operational sub-systems, as illustrated in Fig. 7.5. The number of sub-systems created by key decisions and operational decisions cannot be prescribed for all projects as they will vary considerably in both number and nature depending upon the type of project required and the environmental conditions in which it is to be achieved. However, the model provides a basis for identifying them for individual projects.

Task sub-systems

In arriving at an operational decision, each operational sub-system will consist of at least one task sub-system. The task sub-system level is where the contributors to the project work together to bring forward the propositions upon which operational decisions are based. At operational and key decision level it will be the job of the manager of the process from both the client's organisation and the project team to bring together the decisions in a manner compatible with the client's objectives.

The following illustrates the task sub-system level for an operational sub-system concerned with 'identifying a site' at the same time as which the outline requirements of the client are being finalised and feasibility studies also have to be carried out. This would require three task sub-systems to be operating with reciprocal interdependency. Various people would be carrying out the three tasks and taking advice from a range of different people. For instance, the managing director of the client organisation could be finalising outline requirements with advice from departmental managers and, perhaps, some representatives of the project team; a commercial property agent could be trying to identify suitable sites with advice, say, from the managing director, architect and quantity surveyor; and the feasibility studies would be carried out by the quantity surveyor with advice from the property agent, architect, structural engineer and managing director. The result of this interaction should be propositions which could result in an operational decision to enter negotiations for sites prior to a key decision to purchase a particular site. The problem of managing this type of process is one of integrating such a diverse range of interdependent contributors. In modelling this element in generalised terms, therefore, it is necessary to identify the features that generate the differentiation requiring integration.

The determinants of differentiation have been expressed by Miller (1959) as technology, territory and time and have general relevance to the construction process. The idea of differentiation, as described in the last chapter, provides the tools for analysing different situations that give rise to various levels of differentiation and hence to the need for matching levels of integration. An understanding of the degree of differentiation present in a system significantly aids the person managing the system to provide the appropriate level and mechanisms of integration.

Separation of people working together on a project is affected by the skills (*technology*) they bring to bear on the project. People from different skill backgrounds (e.g. architects, engineer, quantity surveyor, builder) work on problems in different ways, which stem from their experience, and they often cannot see the other contributors' points of view. Such conflicts need reconciling by the manager of the system in terms of the client's objectives. Similarly, separation on the basis of location (*territory*) creates boundaries between contributors. Territory refers to the geographical distance between groups of people working on the project and this obviously affects communication to a very large extent. Often, signboards on building sites will show that the firms contributing to a project are located many miles apart. How much easier it is to resolve a problem on a project through face-to-face discussion rather than by telephone or letter. The advantages to be gained through the contributors working together in the same office (although from different firms), particularly during design, are likely to be significant. A further differentiator is *time*. Although this was visualised by Miller in relation to shift working, it is relevant to the contributors to construction projects in terms of the sequence of activities to be performed where particular contributors cannot perform their activity until another has completed their own.

Overlying and reinforcing differentiation on the basis of the above is sentience, also referred to in the last chapter. It is a particularly strong force in the construction industry and gives rise to strong allegiances to a profession or a firm or to both.

Figure 7.5 illustrates that each operational sub-system will consist of task sub-systems and the people working within them will be differentiated on the basis of skill, location and time, reinforced by sentience. The number and nature of the task sub-systems and the people working within them have to be defined for each individual project as they will vary significantly.

It is perhaps necessary to comment on the relationship of the skill (technology) of project teams and the technological complexity of the project. The skills (technology) needed to design and construct a project are, of course, determined by the technology of the project, that is the skills required for a simple primary school are quite different from those required for the Channel Tunnel. Hence the more technologically complex a project, the greater the differentiation and the need for higher levels of co-ordination (Scott 1992). Bennett (1991) also points out that only when managers and designers have worked out the nature and character of innovative projects can further elements of the project team be added. Therefore, for projects which have standard solutions (i.e. technologically simple), a complete project organisation can be established early but when the technological solution requires much development full teams are established later. Bennett's views on project size are consistent with those on technology. Thus views on the importance of technology and size are consistent with the decision framework described above, as the scale of a project and the difficulty of the technical definition will be reflected in the decision structure. Technology and size give insights to the reasons for certain decisions, and contribute to the nature of the key and operational decisions.

The operating system and the managing system

What has been defined so far in the model is the operating system – that system of activity through which the project is achieved. Figures 7.2 to 7.5 model, in abstract terms, the generality of the construction process. The operating system is managed by a managing system, which carries out the decision-making, maintenance and regulatory activities that keep the operating system going. It is differentiated on the basis of skill from the operating system. The skill of the managing system is management and those of the operating system are professional and technical. The managing system referred to here is that which acts on behalf of the client. It is concerned with the totality of the process of providing the project, which includes that part of the client's organisation relevant to the process. Each system and sub-system into which the process is differentiated may have its own managing system, but such systems will not be managing the total system for the client.

The actual form which the managing system takes in practice will vary considerably. It may be contained solely within the client's organisation, where the client has this capability; in other cases it may consist of a client's representative and a 'consultant' project manager. In the conventional arrangement of contributors it would be the architect acting in a dual role of manager and designer. Alternatively, it may be undertaken by a variety of people at different stages of the project.

The managing system controls the boundaries between the systems and subsystems and integrates their output to ensure that the primary and key decisions made at these boundaries are compatible with the client's requirements. The managing system should ensure that boundaries are appropriately drawn in relation to the process, that facilities for appropriate feedback are available and are used, and that the correct decisions are taken. To achieve this, the managing system also seeks to control the boundaries between the process and its environment, and between the process and the client and its environment.

In order to support this role, the managing system needs to monitor the performance of the systems and sub-systems. Such intra-system regulatory activities are intended to ensure that the manner by which systems and sub-systems arrive at the propositions upon which decisions are based is appropriate. This entails the design and use of feedback mechanisms and requires the managing system to integrate the sub-systems and to ensure that appropriate techniques are used. Although monitoring activities will also be carried out by the systems' and subsystems' own managers, nevertheless the managing system of the total process acting for the client will need to convince itself that the operating system is using appropriate methods.

The managing system also needs to ensure that the resources that produce the output of the systems and sub-systems (for example, and in particular, people) are procured and replenished. These activities aim to ensure that the operating system has the capacity quantitatively to perform its tasks. Such activities will also be carried out by the managers of the systems and sub-systems, but the managing system of the total process acting for the client will again need to determine the fact that the operating system has the capacity to perform its tasks.

Design of organisation structures

The ability of a managing system to operate effectively depends upon an appropriately structured operating system and complimentary managing system. The model has identified, in systems terms, the elements of importance in structuring organisations and has attempted to relate them, in abstract terms, to the construction process. It does not, therefore, present a rigid proposition for the organisation structure of the construction process, but proposes an approach that responds to the specific demands of individual projects. A role of the managing system is to design the organisation through which it will work in seeking to achieve the client's objectives. The manager must, therefore, be provided with the authority to design the operating and managing systems, and to make them function. Such authority will stem from the client, who must decide the pattern of authority established for the project.

Against the background of the model, and accepting that the client's objectives will have been spelled out, an approach to designing an organisation structure for a building project could be:

(a) The manager convinces the client of the need to design an organisation structure for both the construction process and for the client's own organisation's relationship with it, and makes the client aware of the demands that will be placed upon the client's organisation.

(b) The project organisation structure is designed at the very beginning of the process.

(c) Primary decision points are identified.

(d) Within each system created by the primary decision points, the key decision points are anticipated as far as possible.

(e) Within each main sub-system created by the key decision points, operational decision points are identified.

(f) Feedback loops are established within the structure produced by (c), (d), and (e).

(g) The task sub-systems within the organisational sub-systems are identified, together with the skills required to undertake them.

(h) The manner by which the skills are to be provided is established, e.g. separate specialist firms, design-and-build, etc.

(i) Methods of achieving the required level of integration needed because of the differentiation generated by the system are established, including methods of integrating the client.

(j) The pattern of managing system activities is identified as a product of the structure of the operating system and the manner by which skills are provided. This would include the authority and responsibility pattern of the managing system, the client and the contributors to the operating system.

(k) Methods of monitoring, mitigating and harnessing the effects of environmental forces are identified.

The extent to which this approach can be achieved in practice will depend upon the relative certainty of the climate in which the project is being undertaken. With a high level of uncertainty, it may be that the organisation can be designed only a stage at a time, but by going through the process suggested, areas of uncertainty will be recognised. For projects with more stable environments, the organisation structure will be easier to lay down and should present a sound basis for close control of the project.

The essence of the approach is that one starts with a statement of what is to be accomplished through identification of the decision points in the process. Bennett (1991) believes that experienced practitioners can predict key decision points for project types with which they are familiar and at the very least can predict the work required to reach the next key decision point for less familiar projects as there are distinct families of projects with similar sequences of decision points. Identification of decision points is then followed by the design of an operating system required to undertake the tasks prior to each decision, and only then is the managing system designed to suit what is to be achieved. Thus, organisation structure design *follows* the process flow rather than the process having to fit into a pre-determined operating and managing system.

Convergence

The emergence of a theoretical base for project management distinctly separate from the technically based decision theory and operational research schools has seen a convergence which should broaden and enhance understanding of the discipline. Subsequent to the publication of the first edition of this book in 1984 both Bennett's (1985, 1991) and Morris's (1994) work have contributed substantially to the field. Bennett's work, whilst approaching the subject from a different perspective, draws on the same theoretical base, but with particular focus on the work of Mintzberg. His three gestalts (ideal forms of construction project organisation) contribute significantly to project managers thinking about their organisations. Morris's model with three main elements – the definition of the project, its environment and the organisation set up to undertake the organisation – arose from a review of mega-projects in many fields, particularly aerospace, since the 1950s. Whilst Bennett (1994) regrets the lack of an explicit theoretical basis for Morris's analysis he believes that there is a well developed theory running through Morris's thinking. It is significant though that Morris's model emerges from a detailed analysis of real projects rather than the organisation theory literature, particularly as his views appear to coincide with those which arise from this book which is based largely on the literature.

All argue that the choice of an appropriate project organisation is critical to the success of a construction project and provide the theoretical base for the design of such organisations. The project organisation theory which has arisen forms a sound basis on which to understand and design project organisation structures but needs complementing by other branches of organisation theory, such as strategic contingency theory, resource dependency and institutional theory, to provide a comprehensive interpretation of the forces at play on project organisations. Of particular note in this respect will be the transaction cost approach and the resulting organisational economics which are needed for understanding the way in which organisations are created.

Chapter 8
Activities of Project Management

Introduction

The idea of the managing system as separate from the operating system has been introduced previously. The operating system carries out the professional and technical tasks required for a project and the managing system integrates and controls its work. The process being managed starts with the objectives of the client's organisation and transforms them into the aesthetic, functional, time and price criteria for the proposed project and, ultimately, into the completed project itself which should satisfy those criteria. The managing system undertakes a range of activities and fulfils a number of roles in this process.

The range of activities required to be carried out by the operating system varies between projects, depending upon the nature of the project, its environmental context and the consequent arrangement of decision points. The detailed work of the managing system will also vary but, nevertheless, its activities can be conceived irrespective of the structure of the project. They are therefore first discussed in general terms before introducing the more detailed functions of project management.

Managing activities are exercised over the people carrying out each of the tasks making up the project process but more importantly they are concerned with managing the interrelationships of the tasks (or, put another way, with managing the 'space' between people and between tasks), and with managing the relationship of the project to its environment.

Approval and recommendation

Perhaps the most important relationship within the managing system is the connection between the power of approval and the right to make recommendations. The power of approval is, of course, exercised at the decision points in making decisions. The right to make recommendations refers to the authority to make a specific recommendation or to present the alternatives upon which a decision will be based. A person with this role is in a very influential position to persuade the person with approval powers to make a particular decision and so select the alternative that the person with recommendation powers wants.

The managing system normally consists of at least two components, one representing the client and the other the person managing the project for the client. The former will normally be a member of the client's organisation and the latter a member of the project team employed for the specific project. The pattern of approval and recommendation powers between them will depend upon the role the client decides to take and the structure of their organisation. For example, on three projects analysed using the model in this book (Walker & Hughes 1984, 1986, 1987a) clients reserved for themselves approval of the output of most tasks up to commencement of construction, with the exception of a small number of tasks that did not involve choices between alternatives, e.g. preparing contract documentation. The level at which the approval powers were vested in the client organisation's hierarchy depended upon the structure of that organisation. For example, for one project the early decisions were approved by local directors of the parent company until the basic parameters had been established. Then approval powers passed to the client's in-house project engineer. Subsequently, the directors were only involved in approvals at a limited number of decision points. Then, for the construction phase, approval powers passed to the manager of the project team (who was titled project manager and was employed by the engineering consultants). These powers did not, however, include responsibility for approval of further project instructions and of documentation, including drawings, produced by the design team, which the client's project engineer approved. The project manager had the role of recommending courses of action for the client's approval which included presenting and advising upon the choices available. It was this activity from which the project manager's authority on the project was derived. The project manager approved proposals of the contributors, but the final approval to proceed remained with the client.

For two other projects the client's organisation structure was simpler, with the group chairman and managing director respectively representing the client throughout the project; they personally retained approval powers for all decisions during design but not during construction. Again, the manager of the project team recommended actions to the client except during construction, when the manager had approval powers and also had responsibility for approving a small number of routine tasks during design. For these projects additional design information during construction was not subject to approval by the client. The management of the project again drew authority from the power of recommendation.

These arrangements illustrate clients' wishes to be closely associated with their projects and an unwillingness to delegate approval powers. The opportunity for public sector clients to adopt the same position is much more difficult, particularly if the client takes the form of a committee. In this situation the 'client committee' is likely to make one of its officers responsible for managing the project as one of the components of the managing system. This can produce a complex situation regarding approval powers and may result in them being split three ways, part with the 'committee', part with the manager from the client organisation, and part with the manager from the project team.

These issues demonstrate the importance of integration between the client's organisation and the process of construction. It is clear that the client will determine the approval pattern within a project and hence define the authority of its managers of the project. The degree to which this can be formalised will depend upon the will of the client as influenced by the managers. Although it would be ideal to have approval and recommendation powers written down in an explicit form, in reality it is difficult to persuade clients to do this adequately. Nevertheless, it is an ideal for which managers should continue to strive. However, even if it is achieved, there will still remain the problem of interpretation.

The key to these relationships is understanding and communication between the parties involved, which cannot be fully dependent upon formality but which will draw strength from the informal integration of the people concerned through discussion of the issues surrounding the project. In all probability the relationship between approval and recommendation powers will evolve during the project as trust and understanding develop between the participants. Nevertheless, efforts should be made by the client and managers of the project to establish guidelines for these relationships as soon as possible in the project's development.

Boundary control, monitoring and maintenance

Boundary control, monitoring and maintenance are the systems terms which represent the basic project management activities carried out by managing systems. These activities are normally carried out by the component of the managing system drawn from the project teams although if clients' organisations have the capability, they could be carried out by their people.

The objective of boundary control is to ensure functional compatibility of contributors' work within and between tasks, to relate the system to its environment and to control the system's direction towards the required outcome. It is fundamental to the achievement of the level of integration and control demanded by a project and to a satisfactory project outcome. Boundary control derives its name from its role in managing the boundaries between the various sub-systems of the project.

This activity is normally accompanied by the complementary activities of monitoring and maintenance. Monitoring is intra-task regulation to check and control prior to output that a task is proceeding in a manner which will achieve its purpose. Maintenance ensures that a task has the capability to achieve that purpose.

Boundary control involves setting up formal control mechanisms using the feedback loops defined by the key and operational decision points identified when designing the organisation structure, and establishing the supporting information system. It ensures that information flows as intended and that feedback mechanisms are activated. In addition, boundary control should ensure that the reciprocal and sequential interdependencies identified in designing the organisation structure are made to work in the manner intended.

Sequential interdependencies can be integrated by ensuring proper information flow in accordance with the information system, but reciprocal interdependencies need to be integrated using mechanisms that ensure that contributors meet in the correct combinations and at the right times. Such mechanisms would normally include action-minuted meetings and exploratory and less formal meetings in the critical early stages of the project. Where contributors are from different firms, it could, and in many cases should, extend to bringing the people together in one place to work on the project rather than their relying on correspondence and telephone communication. These activities include ensuring that the client is integrated in the appropriate manner at the various stages, and keeping in close contact with the client to identify any changes in environment that may affect the client's requirements. Whereas boundary control relates the parts of the system to each other in the way described earlier, monitoring seeks to ensure that the individuals or groups undertaking a specific task respond to the demands to integrate and also that techniques and procedures appropriate to the specific task are being used.

Maintenance involves keeping in close touch with each contributor and ensuring that each is equipped to carry out the task required. It requires regular formal reviews of the quality and quantity of resources dedicated to the project, particularly in relation to the number and level of skill of the people employed on the project.

Boundary control, monitoring and maintenance are managing system activities and, in accordance with the proposition that the managing and operating systems should be differentiated (on the basis of the skills needed), they should be vested in someone who is not also undertaking operating system activities on the project.

These ideas of what the managing system should be doing sit uncomfortably upon the way in which the construction industry and its professions have evolved. They demand a much greater involvement, some may say interference, by the managing system for the project (e.g. by the project manager) in the activities and tasks carried out by the contributors. The evolution of the industry and its professions has resulted in the creation of a large number of independent firms, but they are interdependent when working together on a project. These ideas, therefore, suggest that control over such firms, both in terms of their relationships with each other and, more significantly, in terms of the activities within a firm, are a legitimate activity of the project management process.

If the managing system is to control the construction process satisfactorily, it requires the authority to carry out boundary control, monitoring and maintenance activities in connection with the activities of all the contributors. The contributors would have to be prepared to accept such authority while still remaining responsible for their individual input, whereas the managing system should be responsible for overall project control. This web of formal relationships is one of the most difficult aspects of structuring a project organisation and one which is not readily faced by clients and project teams. The managing system needs to ensure that it is properly worked out, documented and understood by the participants at an early stage in the process.

Notwithstanding the need for structural relationships to be defined, in reality the effectiveness of the project team will depend ultimately upon the informal relationships generated by the managing system. They will be the product of the way in which it deals with the project team in carrying out its boundary control, monitoring and maintenance activities. Its objectives should be to weld the contributors into a true team and to ensure that they recognise that satisfaction of the client's objectives should be synonymous with satisfaction of their own.

General and direct oversight

There are two further classes of supervision of relevance to the project management level of control, general and direct oversight. Although these are not project management activities as such, the concepts are directly relevant to the effectiveness of the project management process. This does not mean that other managing activities at lower levels in the project hierarchy do not have implications for project management, but they would be the responsibility of the contributing firms and be overseen by the managing system through monitoring and maintenance activities.

General oversight provides policy guidance for the project and *direct oversight* is concerned with directly supervising specific skills used on the project. The manner by which these activities are distributed among the project team depends upon the structure of the firms that contribute to the project organisation. In the case of general oversight (policy guidance), this will often be exercised by the client in the conception process of the project until its broad outlines have been approved. The actual person appointed to undertake this activity depends upon the structure of the client's organisation. If the client's component of the project managing system is provided by a person in authority in the client's organisation, e.g. the managing director, then that person will normally be providing general oversight as well as being part of the managing system. In the early stages of the project, before the project team is involved, that person will be guiding other members of the organisation as the ideas for the development of the project are generated. In such a situation the appointed person may then also become part of the managing system when the project team becomes involved. On the other hand, if the client is represented by a committee, it will provide general oversight. If, as is probable, it delegates to a member of its organisation responsibility for managing the project in conjunction with the project team, the committee will continue to give guidance on policy matters.

A range of possible arrangements are available, and the one selected will be a function of the client's organisation structure. What is important is to understand who is exercising which function.

As the project progresses, the policy guidance will probably pass from the higher management levels of the client's organisation to lower levels. For example, it may pass from the board of directors to the client's 'in-house' project engineer who will become responsible for general oversight as well as being a component in the managing system.

Subsequently, general oversight may pass to the project team when the detailed work on the project commences. The person who then exercises it depends upon the structure of the firm providing the management of the project. For example, if this function is provided by a firm of project managers and the actual project manager is not a partner in the firm, then a partner may provide policy guidance on behalf of the client. The degree to which general oversight on behalf of the client is exercised by someone in this position will depend upon the degree to which the client is prepared to delegate at this stage. If the project manager is also a partner in the firm, he or she will probably exercise both general oversight and project management activities.

A danger at the stage of detailed implementation of the project arises if the client leaves the project team to 'get on with it'. If the managing system is not properly structured and consequently the contributors pursue their work relatively independently, policy guidance may not be provided and the managing system will not be structured in such a way that the lack of guidance is recognised.

The manner by which direct oversight is provided again depends upon the structure of the contributing organisations. Direct oversight is the highest level of supervision exercised over the individual skills used on the project, for example by a partner of one of the contributing professional firms over the activities of people working on the project. Similarly, the relationships between direct supervision and project management depend upon this structure.

For example, where all professional skills and the components of the managing system from the project team are provided by a multidisciplinary firm, direct oversight will be provided by the departmental managers of the firm for each particular skill. Such managers may also be partners of the firm but would not also be acting as project managers if the managing and operating systems are kept separate. Some other member of the firm would be acting as project manager and the relationship between project management and direct oversight in the matrix should be established within the firm. If the contributors are from separate professional firms, their partners will carry out direct oversight, and the relationship with project management, if also provided by a separate firm, will need to be established and could be potentially more difficult to achieve.

Many other arrangements could exist but, again, the important thing is to recognise where responsibilities lie. In many cases, professionally qualified members of contributing firms do not require direct supervision, but this depends upon their status and the policy of the firm by which they are employed.

Because of the frequent use of competitive tenders for construction work and the consequent standard conditions of contract, it is difficult for either general or direct supervision of construction work to be provided by the manager of the project for the client or by any of the design team contributors. This responsibility is vested directly in the main contractor and subcontractors. Standard conditions of contract usually cast the architect or other manager in a passive role in connection with the construction work. The contract is directly between the client and the contractor and the rights and duties of both parties to the contract are specified. The architect or project manager is defined as acting to monitor that the conditions of the contract are carried out. They often cannot intervene directly to ensure compliance with the contract but must follow the administrative procedures laid down, with final recourse to arbitration or law by either party to settle disputes.

If dissatisfied with the contractor's performance, and if satisfaction cannot be achieved by persuasion or mediation, the manager of the project for the client must recommend legal action to the client as a last resort. More purposeful management of the construction stage on behalf of the client will require forms of contract that differ significantly from those commonly used for competitive bidding. This will result in a consequent redistribution of risk and a realignment of responsibility for design and a subsequent redefinition of roles, as found, for example, in management contracting. Other initiatives have sought to tackle such problems in other ways, for example partnering which tries to change attitudes and aid collaboration, and design-and-build contracts which further emphasise the responsibility of the contractor and limit further the scope of the client's representative to manage the construction stage. Whichever approach is selected there remains the need to design an appropriate managing system which ensures that all the contributors respond to the needs of the client.

Pattern of activities

The pattern of managing activities on a project will, therefore, be dependent upon the structure of the firms used in the project organisation and upon the client's organisation structure and the client's requirements regarding the approval powers it wishes to retain. The pattern will also depend to a large extent upon when the client introduces the project team into the process. However, the manager of the project on behalf of the client would normally undertake the activities of boundary control, monitoring and maintenance.

When a project organisation is designed, it is important that the people exercising the various managing activities are identified and their roles understood by all contributors. In this way the authority and responsibilities of the members of the contributing firms will be recognised. For example, it will be known whether the job quantity surveyor has full authority for quantity surveying matters or whether he or she is subject to direct oversight by a more senior member of the firm. This will depend upon the firm from which they come and their status within that firm.

The manager of the project team is usually involved in recommending courses of action to the client for approval. The manager's authority does not therefore generally derive from the power to approve the output of contributors but from the power of recommendation, which implies approval of the output, and hence the power to influence decisions made by the client. The manager's authority stems from access to the client and although this should not bar other contributors from the client if integration of the client is to take place, the latter can vest authority in the manager by considering recommendations only from this source and by requiring other contributors to route recommendations through that manager. Only in this way will the manager have the authority necessary to ensure that the other contributors perform adequately, and have the opportunity to exercise fully his or her integrating activities. Nevertheless, this situation will only be maintained for any length of time if the manager has the professional respect of the other contributors. The manager of the project may be under general supervision by another person higher in the hierarchy of the firm and this may affect the regard in which the manager is held by other contributors, at least initially. The manager's authority is likely to be enhanced if they are a partner or director of the firm.

It would be beneficial if the client were to state formally the authority of the manager of the project and of the other contributors. However, the informal authority of the manager, derived from the respect afforded by the client and other contributors, will be a potent factor and will be the instrument most likely to elicit the necessary level of performance from all contributors.

An example of how activities may be distributed is shown in Fig. 8.1. However, as has been stressed, there are many ways in which roles and activities may be distributed, and Fig. 8.1 shows in outline only one example.

The allocation of responsibility for the project among the contributors will depend upon the association of firms involved and will be the subject of negotiation between the client and the contributors. However, it is possible to put forward ideas as to how responsibility may be distributed. Conditions of contract for construction work will usually define responsibility for this aspect and the related responsibility of the other contributors in connection with this stage, but they do not have to follow the standard format and can be tailored to suit particular projects. Responsibility for design and associated work is the aspect for which responsibility can be more difficult to define.

If a project is managed and designed by a multidisciplinary practice, then responsibility will rest with that firm. Similarly, responsibilities when a conventional arrangement is used, with the architect as designer/manager, are generally understood. In this arrangement, if the consultants are directly appointed by the client, then they will be responsible to the client for their own work. The difficulties that arise in this respect result from the interrelationship of the contributions made and hence from the allocation of final responsibility for specific deficiencies.

If management of the project for the client is given to a firm separate from the firms making up the operating system, a comparable situation will arise if they are appointed directly by the client. Alternatively, if the project is managed by a

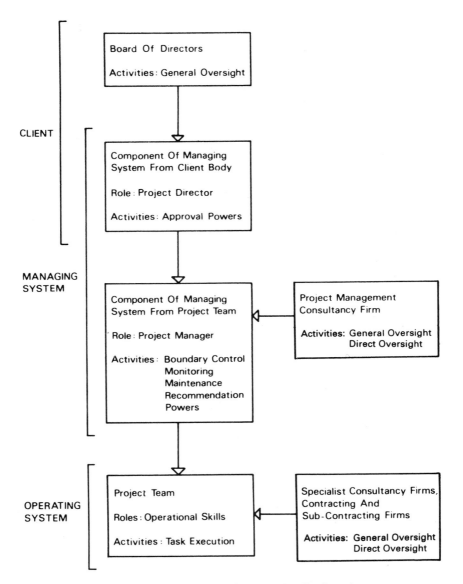

Fig. 8.1 An example of how roles and activities may be distributed.

firm which appoints the consultants directly, i.e. as 'subcontractors' to them, then the managing firm will take responsibility for their work and hence the total project. If a legal action is successfully brought against them by a client, they may have recourse against their 'subcontractors'. This argument can be extended to design-and-build contracts in which one firm will be responsible for the whole of the design and construction of a project.

Naturally, the greater the responsibility accepted by a firm the greater is the risk they are carrying. Firms, whether managing or design-and-build, are unlikely to accept higher responsibility unless they have direct control over the contributors through direct employment, or a facility to bring an action against a contributor if one is brought against them by the client. Importantly, their profit margin would have to be sufficiently high to allow for the level of risk being accepted.

A situation in which responsibility for the project rests with only one firm, or at least one firm for management, design and related aspects and one firm for construction, is likely to be attractive to clients. Informed clients are in a position to dictate the pattern that they want for their project although the reaction of the contributors is likely to be to wish to spread responsibility unless they are appropriately recompensed for risk. Ideally, responsibility should be matched by authority, but this is particularly difficult to achieve. The responsibility pattern adopted should reflect the project structure and the approval and recommendation pattern required by the client. It also should provide the client with legal protection that is sufficiently practical to be applied.

Functions of project management

The abstract account of the activities of project management given previously needs interpreting into the functions that should be expected from project management. In undertaking this, concentration is placed upon project management carried out by the manager of the project team rather than by the client's components of the managing system. This is not to minimise the management of the project that takes place within the client's organisation. It is fundamental to the success of the project and provides the context which to a large extent determines how effective the project team can be.

The client should involve the project team at as early a stage as possible and, prior to involving them, should bring forward the initial need for the project in a co-ordinated and controlled way from within its organisation. If the project team is not involved in the early stages of the project, the client should have taken all useful advice from within its organisation before bringing forward for the project team's advice the strategies that it believed will fulfil the objectives. The client should not have too rigid ideas at this stage as advice from the project team could lead to a better solution to the problem. It would be advantageous if the person who has co-ordinated the client's work and brought forward the strategies could form the client's component of the managing system when the project team is introduced.

The way in which the client carries out this process will be determined by the structure of its organisation but it is important that it should be made explicit to the person from the project team who forms their component of the managing system (referred to in what follows as the project manager). As mentioned previously, the person managing the project team for the client does not have to carry the title of project manager and it may be the architect or engineer who is responsible for managing the process; however, the term is used for convenience.

Consequently, the functions of project management that are discussed here are those that follow from this point and are carried out by the project manager. The detailed functions identified will not necessarily be performed on every project but are intended to illustrate the breadth of project management functions that may be required. Many of them will be repeated in finer degrees of detail at the different stages of the project as further information becomes available. They have not therefore been allocated to the major systems of the process, but require interpretation by the project manager in the light of their degree of applicability at the various stages. The order in which the functions are listed is not intended to indicate a sequence, as many of the functions are interdependent and will overlap in practice.

(1) *Establishment of the client's objectives and priorities*

If the project manager is involved in the conceptual stages of the project, this will be the starting point and it should encompass the client's broader organisational objectives as well as the objectives for the envisaged project. This will enable the project manager to provide informed advice upon the alternative construction strategies available and so provide the basis for developing an appropriate brief for the project.

Even if the project manager is not involved until after the client has arrived at its own conclusions and has some firm ideas about the brief, the client's broader objectives should be explored. This is to make sure that the client's ideas of what the project will provide are sound and are likely to satisfy those objectives.

(2) *Design of the project organisation structure*

The design of the organisation structure of the project should commence by anticipating the decision points in the process and defining the feedback loops and the relationships of the contributors to each other and to the decision points, as described in the last chapter.

(3) *Identification of the way in which the client is integrated into the project*

This will arise from the design of the organisation referred to in (2) above but merits a separate reference. It is important that the project manager persuades the client that the organisation has to be designed to mesh with the project team. Having achieved that, the project manager must ensure that the client responds to the need to integrate with the project team. This will take place at a formal level through meetings but the project manager should seek to ensure that discussions, decisions and the need for those decisions are passed through the appropriate channels of the client's organisation. This will require the project manager to adopt a position close to the client's organisation, and is important for the transmission of information to the client and for the project manager to sense and follow up any changes in the environmental context of the client's organisation that may affect the project.

(4) *Advice on the selection and appointment of the contributors to the project and the establishment of their terms of reference*

The client may require to be advised on these matters. It may even be the case in some circumstances that the client will leave this entirely up to the project manager. It is important that the contributors used on the project have the experience and capability for the particular project and that their approach is compatible with that of the project manager. The project manager is more likely to be able to make this judgement than the client unless the latter is very experienced in construction. The actual terms of appointment of the contributors will, of course, be a decision taken by the client but the project manager should be able to acquaint the client with the alternative approaches available and advise on that which is most suitable for the particular project.

This area also includes the terms of appointment of the project manager and the extent of the project manager's authority. The project manager can hardly advise on whom to appoint as project manager (!), but will have to negotiate the appointment with the client. Perhaps the most difficult decisions the client will have to make are whom to appoint to manage the project and, secondly, how to form an integration with the project manager from within his or her own organisation. The degree of authority delegated to the project manager will depend upon the extent to which the client wishes to retain power of approval, as discussed previously.

(5) *Translation of the client's objectives into a brief for the project team and its transmission*

This involves establishment of user needs, the budget, cost and investment plans and, where appropriate, disposal strategies and their correlation. It is at this stage that fundamental misinterpretations occur and opportunities for economies are overlooked which then become enshrined with the development of the brief. Construction of the brief requires considerable knowledge of the performance in use of buildings, from the points of view of value, maintenance and user activity, and the widest spectrum of advice should be obtained and the most appropriate techniques of evaluation of alternatives employed.

The project manager should make sure that the brief is clearly transmitted to other members of the project team and that it is understood by them. There is a perpetual danger that it will be misinterpreted and result in contributors pulling in different directions. This makes it particularly necessary that it is unequivocally drafted.

(6) *Preparation of the programme for the project*

Although this can be thought of as being part of the brief, it is identified separately since it is an area much neglected, particularly during the early phases of the project and the design stage. The programme should represent a realistic co-ordinated plan of the time needed for the project from the start until, and including, commissioning. As for any plan, it must be carefully monitored, controlled and adapted as necessary.

Although a number of useful techniques exist for programming (e.g. net-working and bar charts), for building work none of the professions involved in the early stages of projects tends to specialise in this work and it is often left to the project manager to provide this operational skill. The situation tends to be different for civil engineering projects for which the engineer will often prepare the programme.

(7) *Activation of the framework of relationships established for the contributors*
Having established the relationships necessary for the project when designing the organisation and agreeing the terms of reference of each contributor, the project manager should ensure that the relationships and responsibilities are activated in the manner intended. The project manager therefore has to be close to the activities of the contributors to ensure that they are in fact performing the work allocated to them and are consulting and taking advice as intended. This activity also covers the work and contribution of the client.

(8) *Establishment of an appropriate information and communication structure*
Much of the information produced for projects is unco-ordinated and this can lead to inefficiency in information use and communication, and to mis-understanding. The project manager should lay down formal communication channels and determine the way in which information is to be presented. This is a particularly difficult problem but the Co-ordinated Project Information system will help and project management firms have devised and developed their own computer-aided data co-ordination systems. In the absence of such sophisticated systems, there is still the opportunity to require the project team to produce information in a compatible manner.

Communication channels present a less intractable problem and the project manager should design them to ensure that all relevant parties are kept up to date formally on events on the project. Of special importance in this respect is communication with the client. The responsibility for preparing co-ordinated reports for the client will rest with the project manager, who should agree the need and frequency of such reports with the client when designing the integrating mechanism between the client and the project team.

(9) *Convening and chairing meetings of appropriate contributors at all stages*
This represents the formal aspects of the integration of the contributors along-side which informal integration will be taking place. Such meetings will act as formal checks on the achievement of the brief in terms of design, cost and time. As such they must be action minuted so that control can be exercised against the decisions of previous meetings. Naturally, minutes should be circulated to all contributors.

It is important that the project manager should be conscious of the need for meetings and adjust their frequency to suit the particular stage of development of the project.

(10) *Monitoring and controlling feasibility studies, design and production to ensure that the brief is being satisfied, including adherence to the budget, investment and programme plans*

This is the 'meat' of the project manager's work and is predominantly concerned with control. The project manager is responsible for taking appropriate action to ensure that the project proceeds to plan. The project manager should be prepared to advise the client if its requirements cannot be met or if an alternative strategy to that contained in the brief emerges as more appropriate to the client's needs.

The project manager will be activating the feedback loops built into the process and measuring progress against the project's objectives, monitoring the project's environment and responding as necessary. The project manager will not therefore be concerned only with the state of the project's development at the time a feedback sample is taken but must also be concerned with forecasting events in the future to anticipate potential problems and attempt to resolve them before they arrive.

Clients generally do not feel that they are well informed about their projects and the responsibility for this lies with the project manager. Thus, as well as involving the client closely in the project, as previously discussed, the project manager should keep the client formally up to date on forecasts of the project team's performance so that action can be taken by the client in advance of a forecast event that may affect its organisation, e.g. delay in completion of the project.

The individual tasks will vary from project to project, but the types of task over which the project manager will have to exercise control and co-ordination are broadly as follows:

- Land acquisition
- Applications for planning consent
- Outline design strategies
- Budget and investment strategies
- Advice on finance, taxation and grants
- Detailed design
- Design cost control
- Disposal strategies
- Proposals for contractual arrangements for construction
- Appointment of the contractor
- Appointment of subcontractors
- Construction
- Cost control during construction
- Disposal

In integrating and controlling the contributors undertaking tasks such as those indicated above, the project manager should ensure that all appropriate contributors are involved in each task and that the output of the tasks are compatible with each other and with the project objectives in terms of design, cost and time.

In order to achieve this, the project manager needs to be assured that the contributors are maintaining an appropriate level of progress in carrying out the tasks and are employing suitable techniques. In addition, the project manager will need to be satisfied that the contributors are using an appropriate number of staff of the right calibre and experience. For this purpose appropriate relationships with the contributors will need to be developed and they will have to accept the project manager's authority to be satisfied on these issues. If the project manager is dissatisfied, authority will be required to ensure that the contributors respond to any reservation held in this respect. This represents a sensitive area and although project managers should be able to achieve their requirements by informal means, they may on occasion have to fall back on an authority given to them by the client.

(11) *Contribution to primary and key decisions and to making operational decisions*
The project manager will bring forward to the client alternative proposals upon which primary and key decisions, as discussed previously, will be based and will assist the client in coming to the decision that best satisfied the objectives. Alternatively, the process may have brought forward a single recommendation. The project manager will undertake the presentation to the client.

Within the process, operational decisions, as discussed previously, will have to be made as a result of the activities of the contributors and it should be the project manager's responsibility to make such decisions.

(12) *Recommendation and control of the implementation of a strategy for disposal or management of the completed project, including commissioning the building and advising on arrangements for running and maintaining it when completed*
This includes the sale or letting of the completed project where appropriate or the 'putting into use' of projects that will be occupied by the owner. In the latter case, maintenance manuals should be provided and 'taught' to the client's personnel.

In commissioning the project, the project manager should involve all the contributors appropriate to this activity so that they can explain to the client's personnel how the project and its services are intended to be used.

(13) *Evaluation of the outcome of the project against its objectives and against interim reports including advice on future strategies*
This represents the final feedback loop and should provide information on the performance of the project team and the client. The distillation of experience on the project should assist the project team and the client to improve their performance on other projects.

Some practical considerations

The Latham Report (Latham 1994) recognises that whilst every project has to be managed, a separate or external project manager may not be required, that many clients have their own in-house project management capability and that the latter will probably be the most satisfactory. Alternatively it sees an external project manager being seconded into the client's organisation. But for the situation in which it is decided to retain an external project manager a series of recommendations are made. These centre on the need to:

- clearly define the appointment and duties of the project manager
- give the project manager the necessary authority
- ensure that fee levels for consultants are sufficient to provide the service required
- interlock the terms of engagement of all consultants.

These are important issues underpinning the effective exercise of the functions previously described. A further major practise matter which is rarely mentioned is the problem of continuity of the project manager. The Latham Report argues for a 'single person/firm', i.e. project manager, to pull the whole process together for the client but recognises that managers have been and still are often appointed for each of the stages of the project, e.g. design, construction. The discontinuity which this approach induces can have serious ramifications for the successful outcome of the project but, even when a project manager is appointed to cover the whole project, the disruption caused by changes in the project manager can have equally deleterious effects on the project outcome. Roberts (1989) believes that when a project has a succession of project managers the project is virtually always adversely affected. As changes of project managers cannot be eliminated, he identified a number of lessons for the incoming project manager:

'• overlap at a milestone whilst the previous manager is still in post
- obtain a list of key contacts and rely less on the project files
- meet key people quickly to clarify key issues
- assume prior decisions were correct
- honour previous commitments
- monitor non-critical path activities, people will have been pursuing diligently those on the critical path but not the others
- if there is an overlap of project managers ensure the outgoing project manager deals with all the less significant administration before he goes.'

He concludes with:

'A final thought that I always keep in mind while in the midst of managing my projects is that I may not finish the project. I know the actions and inactions of my predecessors that caused me the most trouble. I document my files, fulfil promises, update the data base and take care of the non-productive work so that I can pass on projects that truly are clean and without problems.'

An associated problem, identified by Latham (1994) as a reason why there are mixed views about the benefit of project managers in the construction industry in the UK, is that project managers are often appointed after contracts 'run into difficulties or work has already begun and various procurement route options have thereby been blocked off'.

This chapter has intentionally focused on what the project management process should do, not on how project managers (or other titled persons) do it. A later chapter on leadership deals with much about how project managers do their jobs but there is a general level of understanding necessary prior to that, which is briefly discussed here.

The traditional statement of how managers go about their work, arising from the classical approach to management, is that they plan, organise, co-ordinate and control. The great challenge to this view came from Mintzberg (1973) when he asked but what do managers really do? Mintzberg considered that, at best, the classical view indicated some vague objectives managers have when they work. He set about destroying what he felt were four major myths about the manager's job, using the results of systematic research on how managers spend their time.

In terms of planning he found that managers are orientated to action and averse to reflective planning and that their activities are discontinuous with little time to settle on major issues. He found that a manager's work involves regular symbolic duties which cannot be delegated such as being present at ceremonies and negotiations. Also that managers strongly favour verbal communications and that formal information systems play little part in their thinking. He also challenged the view that management was quickly becoming a science and a profession, believing instead that managers rely on judgement and intuition. Instead of what he termed the folklore of management, he identified the basic roles of figurehead, leader, liaisor, environmental monitor, disseminator of information, spokesperson, entrepreneur, disturbance handler and resource allocator. Mintzberg believes that these roles are not easily separable, that they form an integrated whole in which no role can be removed from the framework and the job left intact. Whilst it is not the purpose of this book to develop these ideas in detail, for which the reader is referred to Mintzberg's work, this brief outline is important to provide the backcloth of what project managers (or others) actually do as they carry out project management functions.

The roles identified by Mintzberg and the previous formal statement of the functions of project management both subsume what is probably the most important role of the project manager, that of integrator. If the project manager were unable to fulfil such a role, it is extremely doubtful whether the functions suggested could be carried out satisfactorily.

The list of functions tends to stress the inanimate parts of projects yet one one of the most important components is understanding the human aspects of projects, which can only be achieved by working through people. A vitally important part of the project manager's work is therefore concerned with listening and talking to the members of the project team. This will enable the project manager to anticipate problems that lie at the interface of the work of the contributors, and together with them come up with solutions.

A particularly difficult area in this respect is the project manager's relationship with the contractor and subcontractors when the project has been let by competition. The standard forms and conditions of contract often used in such situations tend, because of the formality and financial implications of the conditions, to inhibit informal relationships between the project manager and design team members, and the contractor and subcontractors. Nevertheless, because of the significance of the construction stage to project success, the project manager needs to be able to establish with the contractor the particular problems anticipated in the construction programme and the action necessary to overcome them. This refers in particular to the transmission of information from the designers to the contractors. It could be the case that, by using other methods of appointing the contractor and subcontractors that allow them to be involved in the design stage, these types of problem will, to a large extent, be overcome.

An essential support to the project management functions is the ability of the project manager to understand other people, to identify what makes them tick and hence to be able to motivate them to perform to the limit of their capabilities. The ability to do this arises from the personal characteristics of the project manager.

Finally, and failing all else, the project manager will need to 'arbitrate' in the case of formalised disputes on the project, whether within the design team, with the construction team or between them, in order to safeguard the client's position. The project manager's knowledge that at some stage such a position may need to be taken will make it more difficult to adopt the integrating role, which is of a more conciliatory nature, but the resolution of such conflict is what is expected of project managers.

Chapter 9
Authority, Power and Politics

Introduction

Authority has been referred to on a number of occasions and in practically every case reference has been to formal authority as the result of a structural position within a project organisation. This simple approach to authority was necessary in order not to deflect attention from the main point being made at that time to which authority was then not central. But now is the time to address the concept of authority in organisations in much more depth and more particularly its more complex companion power and its bedfellow organisational politics. All are strong forces in organisations, be they corporate or project organisations. No matter how well the organisation structure of a project is designed to suit its task and environment, the project will be unsuccessful if the authority pattern is inappropriate and if the powerful players and their political agenda are set against the project. The relationship between authority and power needs to be understood but first a development of the concept of authority used so far is necessary.

Authority

Authority is intrinsic in achieving objectives through organisations. Authority which is vested in a person through the position held in the organisation, and hence the person's right to make decisions upon which others are required to act, is the most common understanding of authority in organisations. Such authority is therefore seen as essential in order to get things done. This simple view of authority describes *formal* or *legal* authority and rests on three basic assumptions (Cleland & King 1972):

- the organisation chart is a realistic descriptive model of an organisation.
- legal (or line) authority is delegated down through the "chain of command". Therefore, if one has legal authority, one can demand the obedience of others.
- given sufficient authority, an individual can accomplish organisational

objectives regardless of the complexity of the forces that are involved (Ries 1964).'

But the concept of formal authority is insufficient in today's complex organisations and informal authority patterns overlay the formal structure. But before moving on to examine such phenomena, the origins of formal authority are worth exploring.

The original source of authority in society can be traced back to private property rights which were held by the crown and the church. They owned the land from which food was produced and hence had the power to enforce their authority. This was the model from which authority in industrial processes arose during the industrial revolution. Independently of the rise of this aspect of traditional management 'theory', Max Weber, the influential German sociologist, produced his typology, (summarised by Scott (1992)) which identifies three types of authority (Weber 1968 trans.):

> '*Traditional authority* – resting on an established belief in the sanctity of immemorial traditions and the legitimacy of those exercising authority under them.
> *Rational–legal authority* – resting on a belief in the "legality" of patterns of normative rules and the right of those elevated to authority under such rules to issue commands.
> *Charismatic authority* – resting on devotion to the specific and exceptional sanctity, heroism or exemplary character of an individual person, and of the normative patterns or order revealed or ordained by him or her.'

Traditional authority has its roots of feudalism, rational–legal authority underpins Weber's notion of bureaucracy. He believed the charismatic forms arose in periods of instability and crisis when people turned to people they believed (rightly or wrongly) could resolve crises. Thus to a large extent a formal view of authority was enshrined within early management thinking.

The work of Barnard (1938) took the understanding of authority significantly forward. His view that goals are imposed from above whilst their achievement depends on willing co-operation from those lower down the organisation, led to his view of authority in which he states that it is a 'fiction that authority comes from above' and that 'the decision as to whether an order has authority or not lies with the person to whom it is addressed and does not reside in persons of authority or those who issue these orders'. These ideas gave root to the concept of *informal* authority which is now explicitly recognised as transcending formal authority and is closely linked to the concept of power which will be discussed later.

These ideas were developed by Dornbusch and Scott (1975) who identified three types of authority: endorsed, authorised and collegial. (They actually referred to these as endorsed power, authorised power and collegial power but the use of power in this context is confusing as explained later when the differences between authority and power are explored.) Originating from Barnard's work they identify authority by endorsement as that arising from a subordinate group acting as a coalition which limit and regulate the exercise of authority over them by a superior. They then go on to recognise that an important characteristic of formal organisations is that each superior will (in the large majority of cases) also have a superior in the hierarchy. Subordinates will have the opportunity to appeal to their superior's superior with the hope that their immediate superior's authority will be curbed. They term the authority of the superior's superior as authority by authorisation. A further source of enforcement of authority norms is seen to be the colleagues or equals of the superior in question. They believe that this source may be expected to be of particular importance in professional organisations and is referred to as authority by collegiate.

Whilst formal (or legal) authority is modified by informal authority much of the management literature tends to overstate the absolute influence of informal authority. In business, people are always likely to respond to a command from their superior. However they may not carry out the order effectively or may seek to dodge it. This latter possibility is high when a subordinate's knowledge of his or her task is greater than that of the superior. Pfeffer (1992), in an account which demonstrates the strength of formal authority, states that 'obedience to authority is conditioned early in life and offers, under most circumstances, many advantages to both society and the individual'.

Whilst legal authority is given by the ability to impose sanctions, the greatest of which in a business sense is the withholding of a salary rise or ultimately dismissal, there are many constraints on legal authority. Formal limitations are contained in laws, contracts, etc., and informal limitations are provided by morality and the capacity of the person so ordered to carry out the task. In fact absolute legal authority does not exist as it depends upon the sanctions being sufficiently high to make the person so commanded obey the order. In a business sense an employee may accept the sanction of dismissal rather than obey the order. In a more extreme example, authority imposed by a state over its people may not be accepted and so revolutions begin and the ultimate sanction of death is not considered great enough to command obedience.

Informal authority is the influence individuals have over the people with whom they interact to the extent that they do as they require. It operates with subordinates, superiors, peers and friends and has to be earned through respect. This type of authority is akin to power depending on the way in which it is used. If misused it can be counter-productive.

Cleland and King (1972) identify the following talents as contributing to an individual's informal authority:

'● superior knowledge
● an ability to persuade people to his way of thinking
● a suitable personality and the ability to establish rapport with others
● a favorable reputation with peers and associates

- a record of accomplishments which lends credence to his or her experience and reputation
- an ability to build confidence in peers and associates
- patience to listen to the problems of subordinates and peers and a willingness to help out when asked or when the need to help is sensed
- an ability to resolve conflict between peers, subordinates, and associates.'

The strength of a person in a position of high formal authority who also has these informal authority characteristics is not difficult to imagine nor are the connections between effective leadership, formal authority and these characteristics.

Power

It is important to distinguish between authority and power in organisations. Power is a much broader concept then authority (Weihrich & Koontz 1993) and many pages have been devoted to its definition. Many stem from Weber's (1947 trans.) definition that 'power is the ability of a person to carry out his own will despite resistance'. Hence 'the ability of individuals or groups to induce or influence the beliefs or actions of other persons or groups' (Weihrich & Koontz 1993), 'the capacity of individual actors to exert their will' (Finkelstein 1992) and the simple definition 'the ability to get things done' (Kanter 1983) used by Lovell (1993) which he believes has a significant appeal in relation to the role of project managers.

Scott (1992) relies on the approach of Emerson (1962)

'It would appear that the power to control or influence the other resides in control over the things he values, which may range all the way from oil resources to ego-support, depending upon the relation in question. In short, power resides implicitly in the other's dependence.'

He believes that Emerson's approach means that power is not to be viewed as a characteristic of an individual but as the property of a social relation. This means that in defining the power of an individual it is necessary to specifying over whom he or she has power. That is, an individual cannot have power generally. Nevertheless, the possession of power is generalised by Pfeffer (1992) in defining power as 'the potential ability to influence behaviour, to change the course of events, to overcome resistance, and to get people to do things that they would not otherwise do'.

Pfeffer defines organisational politics as the exercise or use of power. That is, politics and influence are the processes, actions and behaviours through which this potential power is utilised and realised. Scott comes to a not dissimilar conclusion in combining the definition of power and its use as:

'We will define interpersonal power as the potential for influence that is based on one person's ability and willingness to sanction another person by manipulating rewards and punishments important to the other person.'

Whether such a narrow definition of power is relevant in a construction project management content is open to question and is discussed later.

Relationship between authority and power

Most dictionary definitions of authority contain a reference to power. For example:

'Legal power or right: power derived from office or character or prestige.' (*Chambers Twentieth Century Dictionary*)

We know that formal authority stems from a structural position in an organisation and can be enhanced by informal authority which stems from the characteristics of the person in that position. The person in this position will have power over his or her subordinates which can be conceived as legitimate authority or legitimate power. In this conception the terms authority and power can be seen to be synonymous.

The situation in which power gains its wider meaning is when it is used outside the legitimate (or formal) authority structure such that the person exercising power imposes his or her will on others in the organisation. Such power could be described as illegitimate power as it is used outside the formal organisation structure. For this reason there exists a general feeling of disquiet about its presence in business organisations (it appears much more acceptable in overt political systems.) This leads Pfeffer (1992) to believe that we are ambivalent about power and to quote Kanter (1983) who believes 'Power is America's last dirty word. It is easier to talk about money – and much easier to talk about sex – than it is to talk about power'.

However, in its illegitimate form power can be a 'positive' or 'negative' force. In its positive form those with power will be using it to further objectives which are directly compatible with the organisation's official objectives. In its negative form power will be used to achieve objectives which do not subscribe to organisational objectives. Even in what may appear to be its positive form it can have negative aspects as control of such forces are by definition difficult as they are not explicit and whilst on the face of it may appear positive they may in fact be exerting a negative effect.

Positive power has been described by French and Bell (1990) as 'a balanced pursuit of self-interest and interest in the welfare of others; viewing situations in win–win (non zero-sum) terms as much as possible; engaging in open problem solving and then moving to action and influencing'. They see negative power, on the other hand, as 'an extreme pursuit of self-interest; a tendency to view most situations in win–lose (zero-sum) terms; and predominant use of tactics such as secrecy, surprise, holding hidden agendas, withholding information or deceiving'.

Scott (1992) points out that authority structures are more stable and effective control systems than power structures. The authority structure not only allows a greater measure of control by the person in authority but equally, if not more so, the authority structure also regulates and defines that power. It allows those under authority to act as a coalition in relation to the person in authority to define the limits within which that power can be exercised.

The sources of power

There have been many perspectives on the sources of power in organisations, some of which are complementary, some of which overlap. Together they help to explain the nature of power in organisations.

Emerson (1962) sees power as relational, situational and potentially reciprocal. That is it occurs between people in specific circumstances and may work in both directions. The basis of his formulation is that the power of superiors is based on the sanctions they hold over others and their willingness to use them but that a reward or a penalty is determined by the goals or value of the subordinate in the relationship. In this approach power can have many foundations but writers often refer to power being based on the resources which can be employed in the attainment of desired goals. In this context resources are defined extremely broadly, for example money, skills, knowledge, strength, sex appeal. What types of resource will function as sanctions will vary from one individual to another and from situation to situation (Scott 1992).

Emerson considers that power relations can be reciprocal in that an individual may have power over another in one matter but power may work in the opposite between the same people in another matter. Such power relations are likely to be significant in highly interdependent areas of business such as those found in connection with construction projects.

Within this context there is a reasonably consistent view of the sources of power (Weihrich & Koontz 1993; Newcombe 1994) in the categories of reward, coercive, expert and referent power. Reward power refers to the power to offer enhancement (in many forms), for example, pay, position, conditions of service. Coercive power is closely related to reward power and is the power to punish by way of deprivation of benefits, for example pay, status. Whilst these categories are frequently seen to be legitimate and organisational they do not have to be so. Many indirect opportunities to reward or punish exist outside the formal authority structure and can be seen as the exercise of illegitimate power.

Expert power arises from skill, knowledge and, increasingly, information. Finkelstein (1992) believes that the ability of top managers to deal with environmental contingencies and contribute to organisational success is an important source of power. Managers with relevant experience may have significant influence on a particular strategic choice and are often sought out for their advice. However, power tends to accrue when a manager's expertise is in an area critical to an organisation.

Referent power is the influence that people exercise because people believe in them and their ideas. Finkelstein refers to this as prestige power. He believes that managerial prestige promotes power by facilitating the absorption of uncertainty from the institutional environment both informatically and symbolically. He also believes that prestige also provides power through suggesting that a manager has gilt-edged qualifications and powerful friends.

These sources of power are seen as personal and although they can operate within the formal authority structure, and will have a great effect as confirming its legitimacy, they possess the ability to operate freely outside the legal authority structure and be potent sources of power which can undermine an organisation's objectives if used illegitimately. Scott's definition of power can now be seen to be narrowly drawn and focused on power within the legitimate authority structure of the organisation.

Personality or charisma are often referred to as a source of power but are better perceived as a reinforcing agent to the sources described. The strength gained through expert or referent power will be increased by an appropriate personality as discussed earlier in connection with informal legal authority.

Finkelstein's excellent paper also deals with structural power which is conceived as the equivalent of the formal authority structure. He also identifies ownership power which he defines as 'power accruing to managers in their capacity as agents acting on behalf of shareholders'. He believes that the strength of a manager's position in the agent–principal relationship determines ownership power. The strength of managers' ownership power depends on their ownership position as well as on their links to the founder of a firm so that managers who are also shareholders are more powerful than those who are not.

Power and interdependency

Pfeffer (1992) recognises the link between power and interdependency in organisations. He believes that power is used to different degrees depending on the level of interdependency. With little or no interdependency there is no need for power as there is no situation in which dependency occurs. Similarly he believes that when interdependency is high the motivation to work together is also high and that if this incentive is ignored the organisation is likely to fail. It is in conditions of moderate interdependence that Pfeffer believes that power is more frequently used. However, this view appears to be an oversimplification as it reflects a legal or formal use of power (authority). It is probably the case that generally low interdependency sees little use of power but even in such situations there may be people working on other agendas in which they use illegitimate power with a view to a pay-off in other circumstances in future. This would be much more likely to occur in situations of high interdependency where the opportunity for the use of power is likely to be greater. This situation is one which could be frequently found on complex construction projects.

Interdependency means that dependency of one person or unit on others exists.

It is this dependency which is the basis of power. The ability to diagnose the power structure in organisations is vital to achieving objectives (either legitimate or illegitimate) and, to do this, patterns of interdependency have to be understood. In project management terms this applies within both project teams and client bodies.

Politics in organisations

Organisational politics is the exercise or use of power (Pfeffer 1992). Politics has to do with power, not structure (Mintzberg 1989). Whilst Mintzberg agrees that political activity is to be found in every organisation he believes that politics act to the determinant of the effective functioning of organisations by 'disordering and disintegrating what currently exists'. He unequivocally states that 'I am no fan of politics in organisations' but he recognises that no account of the forces at play in organisations can be complete without a consideration of politics.

Mintzberg uses an analogy of politics as organisational illness which needs to be understood. He sees it working both for and against the system. Politics can undermine healthy processes but can also strengthen a system by acting as a symptom of a more serious disease, enabling early action by provoking the system's adaptive mechanisms. Mintzberg identifies thirteen political games which by his definition involve illegitimate use of power but many of which use legitimate authority as a part of the play. Some co-exist with strong legitimate authority and could not exist without it, others usually highly divisive games are antagonistic to legitimate authority and further games arise when legitimate authority is weak and substitute for it.

Two of the games will be more easily recognised by construction project management people:

> 'Expertise game: non-sanctioned use of expertise to build power base, either by flaunting it or by feigning it; true experts play by exploiting technical skills and knowledge, emphasising the uniqueness, criticality, and irreplaceability of the expertise, also by seeking to keep skills from being programmed, by keeping knowledge to selves; non-experts play by attempting to have their work viewed as expert, ideally to have it declared professional so they alone can control it.

> Rival camps game: played to defeat a rival; typically occurs when alliance or empire-building games result in two major power blocs, giving rise to two-person, zero-sum game in place of n-person game; can be most divisive game of all; conflict can be between units (e.g., between marketing and production in manufacturing firm), between rival personalities, or between two competing missions (as in prisons split between custody and rehabilitation orientations).'

Other games are also relevant, particularly in terms of understanding politics in client organisations and the firms from which project team members come.

As referred to in Chapter 3, Mintzberg has approached the study of organisations on the basis of configurations and identified five basic types of organisations. One type is the political organisation which he describes as:

'What characterises the organisation dominated by politics is a lack of any of the forms of order found in conventional organisations. In other words, the organisation is best described in terms of power, not structure, and that power is exercised in ways not legitimate in conventional organisations. Thus, there is no preferred method of co-ordination, no single dominant part of the organisation, no clear type of decentralisation. Everything depends on the fluidity of informal power, marshalled to win individual issues.'

Of which he identifies four forms:

'• Confrontation: characterised by conflict that is intense, confined, and brief (unstable)
• Shaky alliance: characterised by conflict that is moderate, confined, and possibly enduring (relatively stable)
• Politicised organisation: characterised by conflict that is moderate, pervasive, and possibly enduring (relatively stable, so long as it is sustained by privileged position)
• Complete political arena: characterised by conflict that is intense, pervasive, and brief (unstable).'

The 'complete political arena' appears to be an extreme type of organisation probably rarely found in its absolute form whilst the other three are probably less uncommon but still rare.

The shaky alliance form is the one which is most easily recognised in construction project management terms. It exists when 'two or more major systems of influence or centres of power must co-exist in roughly equal balance'. Mintzberg uses the symphony orchestra to illustrate this form but those in construction will find it familiar, e.g. architect/engineer, designers/constructors.

The political organisation is not the only form subject to political influences which is why it was earlier described as extreme and rarely encountered in its ultimate state. As Mintzberg recognises, politics exist in his other organisation classifications to varying degrees but less extensive than needed to classify them as political organisations.

Two of Mintzberg's organisation categories – professional and innovative – are particularly relevant to the management of construction projects. The professional organisation relies on the standardisation of skills, which is achieved primarily through formal training. It hires professionals for the operating core, then gives them considerable control over their own work. The innovative organisation relies on 'adhocracy'. Mintzberg believes that 'sophisticated innovation requires a very different configuration, one that is able to fuse experts drawn from different disciplines into smoothly functioning ad hoc project teams'. The uniqueness of construction project management organisations is that they often combine Mintzberg's professional and innovative configurations.

What is significant is that Mintzberg believes that there is considerable room for political games in each of these configurations as both have relatively weak systems of authority, though strong systems of expertise. When combined as in construction project management their ability to convert to political organisations must be even greater.

One can do no better than quote Mintzberg:

'The professional configuration may have a relatively stable operating core, where activities are highly standardised, but its administrative structure, where all kinds of professionals and managers interact to make choices, is hardly stable and, in fact, very supportive of power games. The innovative configuration is far less stable, generally having a highly fluid structure throughout that literally promotes games.'

Whilst taking a firmly negative view of politics in organisations Mintzberg does recognise that politics can serve a functional role:

'• As a system of influence it can ensure that the strongest members of an organisation are brought into positions of leadership
• It can ensure that all sides of an issue are fully debated
• It can stimulate change that is blocked by legitimate systems of influence
• It can ease the path for the execution of decisions.'

Whilst reluctant to endorse any political organisation as functional he does acknowledge that a shaky alliance that reflects natural balanced and irreconcilable forces in the organisation could be functional. He uses the example of the differences between research and manufacturing people in a firm which needs the two in balance and considers the alliance functional in such a situation. This is a situation which is directly analogous to design and construction. He goes on to say 'This is because the organisation could not function if it did not accommodate each of these forces. It has no choice but to take the form of a shaky alliance. Some conflict is the inevitable consequence of getting its work done'.

Other views on politics stem from the strategic contingency theorists who argue that power is used to shape the structure of the organisation which reinforces the potential for future political struggle. These views (Hickson *et al.* 1971; Pfeffer 1981) seem less relevant to temporary management organisations then to permanent organisations where long-term positions are likely to pay political dividends.

Power and leadership

The concepts of power and leadership are, of course, closely related. The later chapter on leadership examines the characteristics needed in leaders and the circumstances in which different characteristics and leadership styles are appropriate. Leadership is directly associated with both formal and informal authority which is seen to be legitimate power. The link between leadership and power is therefore that the legitimate power of the leader is the potential for influence held by the leader as a function of his or her formal and informal authority.

This raises the issue of the relationship of illegitimate power which has been a principal focus of this chapter. If strong illegitimate power is being used then it is possible to conceive the idea of an illegitimate leader, that is a leader operating outside the formal authority structure. Whilst there may be some good served by such a situation in terms of compensating for weak or misdirected legitimate leadership, the potential for serious dysfunctional conflict is extremely high. Therefore leadership patterns and performance in organisations should always be considered within the context of the power structure both legitimate and illegitimate.

Empowerment

In dealing with power comprehensively it is necessary to touch on what is seen to be a relatively new phenomenon – empowerment. At first glance it appears to be an elaboration of delegation but close examination shows that it is far more than that. The literature seems reluctant to provide a clear definition but Neilsen's (1986) is useful:

> 'Empowerment is giving subordinates the resources, both psychological and technical, to discover the varieties of power they themselves have and/or accumulate, and therefore which they can use on another's behalf.'

It can be seen that empowerment revolves around the need to provide psychological support and stems from an understanding of powerlessness. According to Kanter (1977):

> 'People held accountable for the results produced by others, whose formal role gives them the right to command but who lack informal political influence, access to resources, outside status, sponsorship, or mobility prospects, are rendered powerless in the organisation . . .'

People who feel powerless do not respond well in the organisational setting, do not provide the level of effectiveness required and often have a negative effect on the achievement of organisational goals. To overcome the negative effects of powerlessness a positive policy of empowerment is necessary. Burke (1986) provides five approaches which allows managers and leaders (between which he draws a distinction) to empower their followers and subordinates as shown in Table 9.1.

Conger and Kanungo (1988) identify leadership practices which are empowering and which are consistent with those of Kanter. They include:

Table 9.1   Differences in the empowering process as a function of role: leaders compared with managers (Burke 1986).

Empowering process	Leaders	Managers
Providing direction for followers/subordinates	Via ideals, vision, a higher purpose, superordinate goals	Via involvement of subordinates in determining paths towards goal accomplishment
Stimulating followers/subordinates	With ideas	With action; things to accomplish
Rewarding followers/subordinates	Informal; personal recognition	Formal; incentive systems
Developing followers/subordinates	By inspiring them to do more than they thought they could do	By involving them in important decision-making activities and providing feedback for potential learning
Appealing to followers/subordinate needs	Appeal to needs of followership and dependency	Appeal to needs for autonomy and independence

'the expression of confidence in subordinates including giving positive emotional support during experiences associated with stress and anxiety, the fostering of opportunities for team members to participate in decision making, the provision of autonomy free from bureaucratic constraint, the observation of others' effectiveness, i.e. providing models of success with which people identify, the setting of inspirational and/or meaningful goals, and, above all, the establishment of a trusting and cooperative culture.'

The benefits of empowerment are claimed to be that it motivates people to face greater challenges that they would if they felt powerless. People are likely to accept higher performance goals and hence leaders are able to put such challenges before subordinates with a reasonable expectation that they will respond. People will be motivated to persist in the face of difficulties due to their increased level of confidence in their own abilities to influence events. Nevertheless, Conger and Kanungo believe that empowerment could have negative effects. They believe that empowerment might lead to overconfidence and hence misjudgements on the part of subordinates. Because of false confidence in positive outcomes, organisations may persist in efforts which are inappropriate to their aims.

The question of responsibility does not seem to be clearly addressed in the literature on empowerment. A key issue would appear to be to ensure that those who are empowered also accept responsibility for their actions otherwise the false confidence referred to by Conger and Kanungo could lead organisations into unwise commitments.

Power in project management

Little has been written about power and political structures, either legitimate or illegitimate, in the project management process yet their use and their potential for influencing the outcome of projects is well recognised by those involved in project management. In his paper on sociological paradigms applied to client briefing, Green (1994) uses a political metaphor to explain the nature of multi-faceted clients. Newcombe (1994) applies a power paradigm to the procurement of construction work contrasting the nature of power in traditional and construction management systems of procurement, Poirot (1991) examined a matrix system and found it to be a power-sharing, power-balancing organisation and Bresnen's (1991) paper purported to show how goal and power differentials can affect project outcomes but, whilst describing how complex construction project organisations are, did not explicitly analyse the power structure.

A comprehensive view of power in projects is given by Lovell (1993) who provides a model of how power affects project managers as shown in Fig 9.1. Lovell is discussing project managers generally and not specifically project managers in construction, hence his use of the term 'Board' in his model. This highlights the distinctive nature of construction project management relative to the more generic concept. The major feature in construction project management, which relates to power structures, is the degree to which the project is handled in-house by the client organisation.

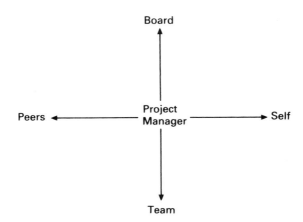

Fig. 9.1 Power and the project manager.
(Reprinted from *International Journal of Project Management*, **11**, Lovell, R.J. 'Power and the project manager', Fig. 1, p. 74, Copyright 1993, with kind permission from Elsevier Science Ltd, The Boulevard, Langford Lane, Kidlington OX5 1GB, UK.)

At one end of the spectrum the project may be totally in-house including all design and construction capabilities, for example for a large developer or a petrochemical company. In such circumstances the potential for the illegitimate use of power would be high as political structures are capable of being established through the long-term relationships which will be present. At the other extreme is the small company which builds very infrequently and hence employs external consultants and perhaps a project manager. In such situations both legitimate and illegitimate power will be of a different kind and scale. Between these two extremes are a whole range of different configurations of client, consultants and contractor. Each project organisation structure will generate different legitimate and illegitimate power structures and the ideas relating to power discussed previously will need to be applied to analyse the appropriateness and effectiveness of the use of power in each case.

Formal and informal authority in projects

The extent to which formal authority is explicitly stated in project structure varies considerably both between projects and within projects. The ultimate in explicitness is contained in the contract for construction work which lays down the rights and obligations of all parties. In the same project it is possible to have the project manager appointed by a brief letter with no explicit statement of either his or her authority or responsibility. Similarly the authority of the client over the project team is invariably implicit. On the other hand the project manager and the consultants may be appointed using formal agreements which are interrelated to show the authority and responsibility of all the contributors.

The formal authority structure within the client organisation itself is of importance to the project team as it will have a significant effect on the establishment of the brief for the project and will determine in the first instance who the project team should be listening to.

Informal authority is a strong force in projects. Project teams are coalitions of highly skilled professionals in which opinions held about the project manager and the other professionals in the team have a profound effect on the informal authority vested in each person in the team. This effect is influenced strongly by sentience, often resulting in the degree of formal authority being substantially modified. Naturally these forces are particularly significant in terms of project success in the way in which they manifest in the project manager. To stand the greatest chance of success the project manager needs strong formal and informal authority.

The influence of the client on informal authority must not be overlooked. Although the client is the highest level of formal authority, any opinion which the client makes explicit about any member of the project team, particularly the project manager, will have a profound effect on that individual's informal authority.

The ideas of endorsed, authorised and collegial authority referred to earlier are a useful way of thinking about authority in projects. The project team can be seen as a coalition which limits the exercise of authority over it by the project manager reflecting endorsed authority. This is due to their collective technical and professional knowledge which the project manager will have to take into account before issuing an instruction with which they may collectively disagree. Authority by authorisation can be seen as the project team's ability to appeal over the head of the project manager to the client. Although unlikely to be used, the fact that the opportunity exists, particularly for a member of the team with strong informal authority, can limit the project manager's absolute authority. Collegial authority is that which operates between equal members of the project team. It is a powerful force in determining which members, and hence specialist contributions, are likely to hold sway when differences in opinion occur.

An issue on construction projects which differs from other organisations is the availability of sanctions. If a project is wholly in-house then the sanctions available to enforce formal authority will be similar to those in other organisations. However, when elements of the project team are external to the client organisation the available sanctions are different in nature, or at least the likelihood of them being used is remote. It would be a dramatic and potentially disastrous course of action to fire any of the consultancy firms. Whilst this has occurred on some projects it would be an absolute last resort. Intermediate sanctions are not readily available other than a threat to withhold future commissions.

Power in projects

The meaning of power used here is its use outside the legitimate or formal authority structure of the project. As previously stated, its potential for use will be much greater on wholly in-house projects and may well operate in such situations in ways which are comparable with the way in which it operates in organisations generally, as described in the mainstream management literature on power. It is in circumstances where projects are not carried out wholly in-house where special features of power arise specifically related to the organisation structure of construction projects.

The exercise of power in client organisations can have major effects on a project's objectives and how they are defined. Power struggles within client organisations can distort objectives leading to the construction of a project which the client considers unsatisfactory on completion (Cherns & Bryant 1984). The reason for dissatisfaction will stem from how the client initially defined the project (often as a result of a power-driven political process within the client organisation) but this will not stop the client believing that it was due to some fault on the part of the project team. Such an outcome can affect the credibility of the project team such that the strength of their informal authority is reduced and so may create problems for them on future projects. This effect does not apply only to the project manager but also to individual members of the project team.

An analysis of the construction of a new university showed that the exercise of power within a complex client structure and a sensitive wider political context meant that the project was commenced without a full understanding of the type of university to be constructed (Walker, A. 1994). Whilst the project was completed extremely quickly the final cost was three times the original estimate. Nevertheless the project was considered to be good value for money. However, the members of the project team were perceived to have created 'a massive cost overrun' and suffered from this reputation resulting in some of them being rejected for future commissions. The 'overrun' was all that was remembered. The final cost would not have been seen as an overrun if the power plays within the political context had not resulted in a commitment to an extremely short period for design and construction and the resulting lack of cost management.

Power can, of course, work for good and bad. Whilst the above illustration shows that the outcome was negative for the consultants, the exercise of power was positive in delivering a new university three years earlier than expected to the benefit of the community as a whole. The potential for the use of power within public client organisations is probably even greater than for private sector projects. The Hong Kong Convention and Exhibition Centre took ten years from conception to the start of design due to competing power plays by government departments. Only when a chairperson held in very high regard in the community and reinforced with charisma was appointed to the Trade Development Council (TDC) was sufficient power vested in the TDC to enable it to break the deadlock (Walker & Kalinowski 1994).

A distinctive power source on construction projects is that of patronage. This stems from the process of obtaining commissions for future projects. Whilst increasingly commissions are awarded on the basis of competitive bids, particularly in the public sector, nevertheless in the private sector commissions are frequently awarded on the basis of recommendations. Even for competitive bids, recommendations are often needed in order to be admitted to the list of bidders. Traditionally architects had the gift of selecting the consultants with whom they would work. In contemporary times, this gift has in many cases moved to the project manager. This situation gives the project manager or the architect enormous power over the consultants as they will be unwilling to go against the desires of those who may hold the gift of future work. Even on projects where consultants have been appointed in competition the position is no different as the process may introduce consultants to project managers or architects with whom they have not previously worked. The potential of future commissions from new sources will act to give power to the project manager or architect.

The sources of power described earlier (reward, coercive, expert, referent) need relating to construction. In the management literature reward and coercive power are seen to be legitimate power (or authority) as they offer benefits or sanctions which operate within the formal structure. However it has been argued above that this may not be the case for construction projects. Rewards and coercion can be applied by those outside the formal authority structure of the project and may thus distort the advice and actions taken by the dependent party where such advice and action is opposed to the view of the power holder yet is in the best interests of achieving the project's objectives. Project managers need to be very sensitive to such issues.

Expert power is a major force on construction projects. Projects are so technologically and managerially demanding that reputation and track record count for a great deal. The specialisation of professional skills also contributes to the effectiveness of expertise as a power base. Project managers and others have great difficulty in contradicting the expert advice of a contributor even though they may instinctively believe that it is inappropriate. A corollary is that such experts are likely to gather such support from colleagues against less expert members of the team who may disagree with the expert.

Referent (or prestige) power is the influence that people exercise because people believe in them. Although closely related to expertise power it is more associated with a record of substantial achievement. This source of power is particularly reinforced by charisma. It is a particularly potent force on projects at a number of levels. At the initiation of projects people with this type of power are able to force through the acceptance of projects 'against the odds'. For example the selection of one airport site as opposed to another. Project managers with referent power are naturally in a strong position as are other members of the project team, for example an architect with an international reputation for the design of famous buildings. Project managers and others who worked on the Mass Transit Railway in Hong Kong, which was an extremely successful project, gained great referent power and went on to obtain other important project management and other posts as a result.

At another level, a play which relies on referent power is when one of the consultancy firms sends along a director to a project meeting which would normally be attended by a project level staff member, for the purpose of obtaining a particular outcome. The intimidation of other members of the meeting by the presence of a director can be seen as the use of referent power.

Architects as a profession gain referent power from their historical social position. This certainly still occurs in the UK, though probably less than in the past, and no doubt in other countries. In the early part of this century architects emerged as the elite social class associated with construction stemming from the patronage of their clients as described by Bowley (1966). Architects continue to derive referent power from this historical context.

Finkelstein (1992) refers, in passing, to prestige power also being provided through a manager having powerful friends. This facet assumes a more significant position in construction. Due to the high interdependence of the elements of the process both inside and outside the project management system, the use of 'powerful friends', or rather better put as 'contacts with influence', plays a large part in moving the project forward at all stages. Such influence can include, for example, arranging finance at the early stages, contacts with government to ease the approvals process and good contacts with suppliers and subcontractors during construction. People who can get things done through their contacts acquire strong prestige or referent power.

Politics, projects and firms

Previously construction project organisations were classified (using Mintzberg's taxonomy) as shaky alliances of professional and innovative organisations. The shaky alliance being one of Mintzberg's types of political organisation.

It is worth continuing this discussion at this point. Firstly, this configuration applies to projects for which the project team is not wholly in-house. If the project team was wholly in-house it could be part of a political organisation which fitted any of the other three forms – confrontational, politicised or complete political arena. In such circumstances the development of an effective project could be difficult to achieve although perhaps less so in a politicised organisation. A project team not wholly in-house would be unlikely to fit such descriptions.

However, even when not wholly in-house, these typologies have relevance for understanding the effect of organisational politics on projects. The client organisation could be of any of these forms at the time the project is being initiated and objectives are being defined. The possibility of obtaining a suitable brief must be extremely limited in the case of the confrontational and complete political arena forms but the shaky alliance and politicised organisation should be less difficult if understood by the project manager and project team.

The other important elements which have the potential to become political organisations are the individual consultancy firms which make up the project team. Mintzberg would categorise them as professional organisations which he believes is one of the organisational types in which there is considerable room for political games. If so, they have the propensity to become political organisations of a confrontational, politicised or complete political arena type. If a consultancy firm transformed from a professional organisation to a political organisation during the course of a project, the effect on the project could be severe. The likelihood would be that rather than focusing on the project, the attention of members of the firm allocated to the project would be deflected to the internal problems of the firm. In an activity as closely interdependent as a construction project this could be to the detriment of the project.

Empowerment and projects

Empowerment is a relatively new buzz-word in management and is important to many organisations which include powerless and disadvantaged groups. However, in the process of managing construction projects empowerment of the project team occurs naturally due to the nature of the task which requires highly skilled professionals to be self-motivated and accept high performance goals. This they have always done due to their professional training and traditions so the concept is not new although it has not been identified and given the name empowerment.

Generally speaking project managers instinctively empower their project teams although little effort is needed as the professional contributors generally empower themselves. Nevertheless this may not occur on every occasion so the concept should be understood and applied by project managers to further enhance the project team's contribution.

However, if the construction team is included in this scenario, Newcombe (1994) claims that the construction management procurement system produces a different power configuration than the traditional procurement system. He believes this system allows a better opportunity for the exercise of expertise power and charisma coupled with the empowerment of all the parties especially the specialist trades contractors, but empirical work is required to substantiate these views.

Project managers and power

Project managers need to recognise the power configurations in projects. They need to recognise that power has a potential for influence over others outside the legitimate authority structure. If such influences are seen to be harmful to achieving project objectives, the project manager needs to work out strategies to deal with them. At the same time the project manager needs to avoid becoming paranoid over power which is easily done when working in complex organisations. Power plays, if they exist, need turning to positive not negative ends.

Pfeffer (1992) has constructed some questions to be used by managers for understanding the pattern of power in organisations which are also useful for project managers:

'• Decide what your goals are, what you are trying to accomplish.
  • Diagnose patterns of dependence and interdependence; what individuals are influential and important in you achieving your goal?
  • What are their points of view likely to be? How will they feel about what you are trying to do?
  • What are their power bases? Which of them is more influential in the decision?
  • What are your bases of power and influence? What bases of influence can you develop to gain more control over the situation?
  • Which of the various strategies and tactics for exercising power seem most appropriate and are likely to be effective, given the situation you confront?
  • Based on the above, choose a course of action to get something done.'

Chapter 10
Project Leadership

Introduction

In the first chapter care was taken to distinguish between the title 'project manager' and the process of 'project management'. Subsequently focus has been principally upon the process with relatively infrequent references to the role of the project manager. The last two chapters have looked at the activities of project management and authority in projects so that it is now appropriate to consider certain aspects relevant to the person who is charged with making the organisation work effectively. For convenience this person will be referred to as the project manager, but may go under a number of titles in practice, and will refer to the person managing the project for the client.

This chapter will not set itself the impossible task of describing the perfect project manager – that holy grail that all should seek. Rather it will examine the major aspect of performance of project managers which aids effective management – leadership.

Leadership is an intrinsic part of management. It is the manner in which managers conduct themselves in their role in order to obtain the best performance from the people they are managing. Hence all managers lead, some much more effectively than others. It takes many forms depending on the characteristics of the leader and the demands of the situation in which it is exercised. It is rarely needed in the form of the public perception of it – the 'knight on a white charger' – but much more frequently as perceived by the Chinese philosopher Lao-tzu – 'to lead the people, walk behind them'.

An understanding of the way in which the actual application of leadership skills to project teams takes place is necessary for project managers and this is examined, followed by a consideration of the formal context of the project manager's position. This involves an understanding of terms of reference for the appointment of project managers and the legal liabilities which arise from such appointments.

Leadership

Much has been written and a number of major research studies have been undertaken on leadership, without a convincing outcome. It is generally accepted that leadership is important for management to be effective and this tends to create a general perception that management and leadership are somehow separate. Management is seen to be a mechanical process using techniques, responding to directives from elsewhere and controlling those being managed. Leadership is perceived as charismatic, inspirational and forward thinking. However, the definition of management used in this book includes the phrase 'working through others'. Immediately this is recognised, leadership and management become intrinsically linked.

The public perception of management, as a mechanistic process, is better described as administration. Townsend (1984) believes that 'most people in big companies today are administered not led. They are treated as personnel, not people' and he quotes The *Peter Principle* (Peter & Hull 1969) in support, 'most hierarchies are nowadays so cumbered with rules and traditions and so bound in by public laws, that even high employees do not have to lead anyone anywhere, in the sense of pointing out the direction and setting the pace. They simply follow precedents, obey regulations, and move at the head of the crowd. Such employees lead only in the sense that the carved wooden figurehead leads the ship.'

Leader and manager cannot be separated. The quality of a manager's leadership will be demonstrated by the quality of achievement of the people he or she leads. The dangers of having a project administrator rather than a project manager or project leader are all too apparent from the foregoing. Project managers should lead not administer project teams if they are to be effective in delivering projects which are functionally effective, on time and within budget. But this does not have to be high profile, charismatic leadership, it may be much more effective in another of Lao-tzu's conceptions, 'As for the best leaders, the people do not notice their existence. The next best, the people honour and praise. The next, the people fear; and the next the people hate ... When the best leader's work is done the people say "we did it ourselves".'

Some research models

It used to be believed that leaders were born not made, a myth no doubt encouraged by the leaders themselves. Whilst research has shown that leaders are not born as such, it has not been able to show clearly how leaders do emerge. Certainly personal characteristics do contribute to the ability of people to lead but personality develops through experience and is only partly hereditary.

Nor is there agreement on the essential qualities of a leader. This is even so in the military field, which has a well developed sense of leadership in its terms. The following lists of qualities of leadership from four military establishments illustrate the point:

Navy	*Air Force*	*Army*	*Marines*
Faith	Efficiency	Bearing	Integrity
Courage	Energy	Courage	Knowledge
Loyalty	Sympathy	(Physical and	Courage
Sense of duty	Resolution	moral)	Decisiveness
Integrity	Courage	Decisiveness	Dependability
Humanity	Tenacity	Endurance	Initiative
Commonsense	Personality	Initiative	Tact
Good judgement		Integrity	Justice
Tenacity		Judgement	Enthusiasm
Fortitude		Justice	Bearing
Physical fitness		Loyalty	Endurance
Mental fitness		Tact	Unselfishness
Self-control		Unselfishness	Loyalty
Cheerfulness			Judgement
Knowledge			

If the military have such variety, it can be seen that the difficulties facing researchers into leadership in business and industry are enormous.

As mentioned earlier, research has not been able to show clearly how leaders do emerge, nevertheless, some valuable directions and tendencies have been identified which are useful to managers who are seeking to improve their leadership skills.

Much leadership research stemmed from the work of Fielder (1967) in the late 1960s. His model argues for a contingency-orientated approach in which two factors were particularly important:

(a) the degree to which the leader is liked and trusted by the group
(b) the degree to which the group's work is defined.

The contingency approach is best illustrated by polarising two styles of leader:

(a) the directive managing, task-orientated leader
(b) the non-directive human relations-orientated leader.

Fielder argued that each style was appropriate in particular situations. In very favourable situations, i.e. when the leader is well liked and the group's work is clearly defined, and in very unfavourable situations, i.e. the opposite of the last, a directive management approach is better. Conversely if the leader was well liked but the group's work unstructured and when the leader is not liked but the group's work is clearly defined then a non-directive approach would be beneficial. This is a very severe simplification but gives a sense of the developments in thinking about leadership which was engendered by Fielder.

This work had been preceded by Likert's which concluded that the best supervisors were employee centred. The range of styles implied by Fielder's work placed Likert's into perspective by recognising a directive management approach in certain circumstances.

The development of the idea of the usefulness of a range of leadership styles, each to be used in the appropriate setting, was demonstrated by Tannenbaum and Schmidt (1973). A simplified range is given in Fig. 10.1.

This work implies that the leader needs great sensitivity to the situation regarding the feeling of the group being led and the nature of the task being undertaken and needs to be sufficiently flexible to change styles as appropriate.

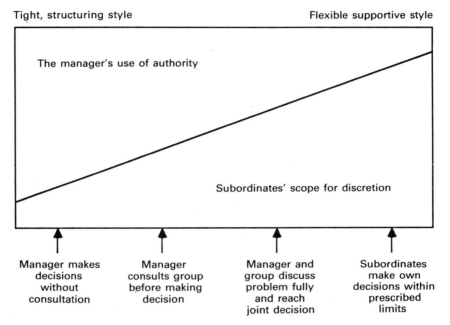

Tight, structuring style   Flexible supportive style

The manager's use of authority

Subordinates' scope for discretion

| Manager makes decisions without consultation | Manager consults group before making decision | Manager and group discuss problem fully and reach joint decision | Subordinates make own decisions within prescribed limits |

Fig. 10.1   Some styles of leadership (adapted from Tannenbaum & Schmidt 1973 by Fryer 1985).

This approach is further illustrated by the work of Hersey and Blanchard (1972) and of Blake and Mouton (1978) who considered leadership styles as an amalgam of task and relationship-orientated approaches (equivalent to the directive and non-directive styles referred to earlier). The appropriate mixture to be used should be determined by the maturity of the group being led and the needs of the situation. Maturity is defined in terms of achievement, motivation, the willingness and ability to take responsibility, education and experience. It is suggested that leadership behaviour should relate to high/low task and high/low relationship permutations as illustrated in Table 10.1. The relevance of these permutations to the maturity of the group being led is illustrated in Fig. 10.2.

Bonoma and Slevin (1978) have used information input and decision authority in formulating their graphical model of leadership. They believe that the leader's need for information from the group he is leading and also where the leader allows the decision authority to lie, are the factors which determine the leadership style to be adopted. Their graphical model is illustrated in Fig. 10.3.

Table 10.1   Basic leader behaviour styles. (*Sources:* Hersey & Blanchard 1972; Blake & Mouton 1978; Gilbert 1983).

Basic styles	Situation	Effective leadership	Ineffective leadership
High task and low relationships	Efficiency in operations as a result of arranging conditions of work in such a way that human elements interfere to a minimum degree	Seen as knowing what he wants and imposing his methods for accomplishing this without creating resentment.	Seen as having no confidence in others, unpleasant and interested only in short-term output.
High task and high relationships	Work accomplishment is from committed people, whose interdependence, through a 'common stake' in organisation purpose leads to relationships of trust and respect	Seen as satisfying the needs of the group for setting goals and organising work but also providing high levels of socio-emotional support	Seen as initiating more structure than is needed by the group and spends more time on socio-emotional support than necessary
High relationships and low task	Thoughtful attention to the needs of people for satisfying relationships leads to a comfortable friendly organisation atmosphere and work tempo	Seen as having implicit trust in people and as being primarily concerned with developing their talents	Seen as primarily interested in harmony and being seen as 'a good person' and unwilling to risk disruption of a relationship to accomplish a task.
Low task and low relationships	Exertion of the minimum leadership effort to accomplish the required work is appropriate to sustain organisation membership.	Seen as appropriately permitting his subordinates to decide how the work should be done and playing only a minor part in their social interaction	Seen as uninvolved and passive, as a 'paper shuffler', who cares little about the task at hand or the people involved

As Slevin (1983) states:

'Using this plotting system, we can describe almost any leadership style. However, the four extremes of leaders you have known (depicted in the four corners of the grid) are the following:

(1) Autocrat (10, 0). Such managers solicit little or no information input from their group and make the managerial decision solely by themselves.

(2) Consultative Autocrat (10, 10). In this managerial style intensive information input is elicited from the members, but such formal leaders keep all substantive decision-making authority to themselves.

(3) Consensus Manager (0, 10). Purely consensual managers throw open the problem to the group for discussion (information input) and simultaneously allow or encourage the entire group to make the relevant decision.

(4) Shareholder Manager (0, 10). This position is literally poor management. Little or no information input and exchange takes place within the group context, while the group itself is provided ultimate authority for the final decision.'

Given the four positions in the Bonoma–Slevin model, the leaders should be able to move their style around the graph in response to characteristics of themselves and the situation they face. Bonoma and Slevin (1978) have tabulated the types of pressures that leaders might face and the directions that these pressures might push them. Using the terminology North, South, East and West, you can look at the graph and follow the direction that each pressure might tend to move the leader (see Table 10.2).

More recent work has linked leadership style to organisation culture (Quinn 1988) in that leaders should function in a manner consistent with the desired culture of the organisation. However, whilst there exist many definitions of organisation culture, there is not yet an accepted approach to research into organisation culture so its relationship to leadership styles is not yet established (Maloney & Federle 1991).

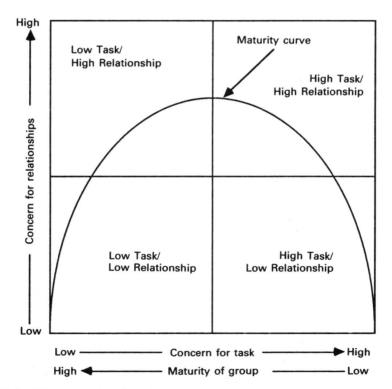

Fig. 10.2 Effective leadership styles versus group maturity. (*Sources:* Hersey & Blanchard 1972; Blake & Mouton 1978; Gilbert 1983.)

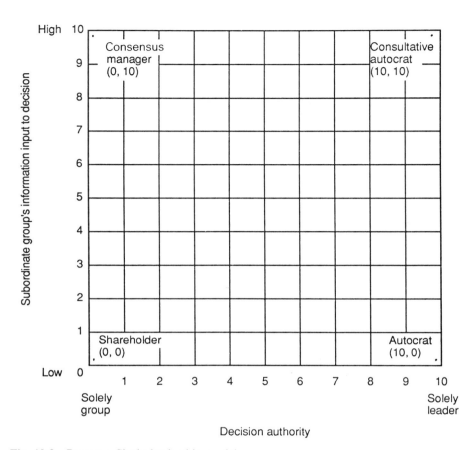

Fig. 10.3   Bonoma–Slevin leadership model.

The key to successful leadership is knowing what your dominant style is and being able to modify that style depending upon the contingencies of the various leadership situations that you face, which are likely to include your organisation's culture.

Leadership style

One thing in common between the researchers into leadership is the recognition of different management styles along a spectrum from authoritarian to democratic, although different terms are sometimes used, for example 'task-centred' or 'employee or people (relationship) centred'. At the extreme ends of the spectrum the authoritarian manager issues orders without consultation and the democratic manager allows the group to take the decision after having fully discussed the

Table 10.2 Three leadership style pressures. (*Source:* Bonoma, Thomas V. and Slevin, Dennis P. *Executive Survival Manual*, Wadsworth Publishing Company, Belmont, CA, 1978, p.89.)

1. PROBLEM ATTRIBUTE PRESSURES	DIRECTION OF PRESSURE ON LEADERSHIP GRID
• Leader lacks relevant information; problem is ambiguous.	North – more information needed.
• Leader lacks enough time to make decision adequately.	South and east – consensus and information collection take time.
• Decision is important or critical to organization.	North – information search maximized.
• Decision is personally important to leader.	North and east – personal control and information maximized.
• Problem is structured or routine.	South and east – little time as possible spent on decision.
• Decision implementation by subordinates is critical to success.	West and north – input and consensus required.

2. LEADER PERSONALITY PRESSURES	
• Leader has high need for power.	East – personal control maximized.
• Leader is high in need for affiliation or is "people oriented".	North and west – contact with people maximized.
• Leader is highly intelligent	East – personal competence demonstrated.
• Leader has high need for achievement.	East – personal contribution maximized.

3. ORGANIZATION/GROUP PRESSURES	
• Conflict is likely to result from the decision.	North and west – "participative aspects" of decision making maximized.
• Good leader-group relations exist.	North and west – group contact maximized.
• Centrality, formalization of organization is high.	South and east – organization style matched.

issues. Much of the research suggests that the situation to be managed should determine at which point on the spectrum the leader should be.

The practical application of these ideas does create problems. They presuppose that the leader's leader has the ability to determine the type of leader needed in each situation and to select such a person. This also implies that either if the situation changes the leader should be changed or the leader should change style if the situation changes. The latter seems to be favoured by the writers but it is extremely difficult for an individual to change management styles and for the group, assuming it does not change, to accept different behaviour styles from the same leader. By the time someone reaches a high management level, their character and personality will already have been established and the manager is likely to have settled in a position on the autocratic/democratic spectrum and have a relatively short span of flexibility. If a wide range of flexibility is to be expected, this has major implications for the way in which both project managers and project teams are trained for the construction industry as this degree of flexibility does not exist at present.

Leadership and the project manager

As discussed, there is no definitive research on leadership. What the research has shown is that there are a large number of variables which contribute to determining the style of leadership needed in a given situation in order to manage effectively. They include whether the leader is liked, how closely the work of the group is defined, the maturity of the group being managed, the location of decision authority and the degree of information input. The research reviewed is only a sample of the research which has been undertaken on leadership, hence the variables listed are by no means exhaustive. They are used to illustrate the sort of features that managers should consider when thinking about the effectiveness of their leadership.

Research on leadership has focused on general management situations not on project management. It has already been illustrated that the management of a construction project has unique features such as the relationship with the client and the interorganisational nature of the process. The project manager is usually in the position of leading contributors from a large number of commercially independent organisations over whom he has only limited formal authority. A significant outcome of this situation is that, in the matrix, each contributor will be subject to 'leadership' from both the project manager and the manager of the employer's organisation. In effect the 'leaders' are in 'competition'. As long as the leaders have the same objectives there should be no problems but the potential for the contributor's leader from the employing organisation to distort the project manager's leadership is present.

Project managers will be leading a group of mature, experienced professionals and are often only slightly elevated over them in terms of legitimate authority. They may even resent having project managers in this position. Consequently the project manager's leadership style will tend to be democratic and rely on influence and persuasion rather than formal authority. Project managers are continually in receipt of information from their teams and must interact with them, constantly exchanging large volumes of information of a technical and financial nature. A not dissimilar relationship exists with the client with the important exception that the client has final formal authority over the project manager. Nevertheless a similar style, based on persuasion, will usually be appropriate for the project manager, although it can be seen as a 'subordinate' leading a 'superior'.

Each of the sub-systems in the construction process terminates in either a key or operational decision. The nature of the work to be undertaken within each sub-system will determine the leadership style most appropriate to arriving at the best outcome. However in each case the decision will effectively be made by the project manager rather than the project team, although in the case of key deci-sions it will be in the form of a recommendation to the client. So whilst for some sub-systems the project manager should adopt a democratic form of leadership as ideas are formed and developed, at the end of the process it will be the project manager acting 'autocratically' who will make or recommend a decision. It will be the project manager's responsibility to do so although in practice it may not appear to be autocratic as in many cases the decision will be arrived at by con-sensus. However, many project managers may not wish to appear in such a high profile decision-making role and may seek to have decisions seen as group decisions. There are degrees of democracy and the actual position on the scale taken by project managers will depend on a number of factors such as the urgency of the decision, their personal characteristics and the experience of the group being led. The sub-systems which lend themselves to a democratic style of lea-dership are those concerned with developing initial evaluations, outline design strategies and selecting contractual arrangements.

Other sub-systems may lend themselves to a leadership style which is more towards the autocratic end of the scale. These will include, developing working drawings, preparing contract documentation, cost control during construction and many during the construction stage if let by conventional competitive tender. However, should problems occur on any such mechanistic sub-systems it may be that a more democratic style of leadership may be necessary to enable the group to solve the problem. Conversely within a sub-system being led democratically it may be necessary on occasions to be autocratic to bring it to a conclusion and make progress.

The project manager has the taxing job of playing the scales from democratic to autocratic leadership depending on the needs of the particular sub-system or the particular stage of development of the sub-system. Anderson (1992) con-firmed this view and found that leadership skills had the highest frequency of significance across eight project management functions examined yet project managers were perceived as having only average or less than average managerial skills. However, it should be recognised that the project manager's need for flexibility is extremely high not only because of the nature of the process being managed but also because of the maturity and organisational independence of the contributors who originate from separate firms. In such a context the project manager's position on the scale will lay predominantly towards the democratic end as he or she seeks to cope with what is inevitably a challenging leadership position.

Whilst the demands of project management may be rather different from general management, it will be helpful to project managers for them to review the previously described research models. This will enable them to ask themselves questions about how they lead project teams, and help them to focus on the variables which have been seen to have an effect on leadership.

Fielder's (1967) work makes a good beginning by asking project managers to consider how well liked they are by their teams and also to evaluate the degree of uncertainty of the task being undertaken. The latter is very variable in construction, ranging often from practically 'a blank sheet of paper' in the early stages of some projects to what can be a well defined task of constructing the building. In contrast for some standard building types the degree of certainty of the tasks is very high. It also asks project managers to assess where they see themselves on a scale from autocratic to democratic manager and the degree of flexibility which they consider themselves to have. Unfortunately these are not questions which we can easily answer about ourselves nor are they answers we like to hear from others. Herein lies one of the greatest dilemmas in improving leadership abilities – that of recognising honestly one's own abilities.

The flexibility of leadership styles needed by project managers is clearly demonstrated by Tannenbaum and Schmidt's scale (Fig. 10.1). It is clear that the wide range of situations encountered on construction projects will require styles covering much of the scale, although there will be a strong bias towards the right hand end due to the appropriateness of professionals making decisions within their competence. This begins to illustrate that project managers will need to have a fundamentally democratic style of management.

The low task and low relationship leadership style of Blake and Mouton is clearly appropriate to many construction projects for which the project manager is dealing with highly educated, well trained and emotionally mature people. It also demonstrates other modes which may be necessary in specific circumstances, particularly if a crisis situation occurs. However, such situations are rare and the project manager is more likely to have to switch to a low task/high relationship, or even a high task/high relationship mode, if leading a somewhat less mature group particularly in times of high construction activity when less experienced staff occupy more senior positions.

The range of leadership styles is well illustrated in Bonoma and Slevin's model (Fig. 10.3) and is particularly useful for project managers who are dependent upon their team of professionals to provide them with information and who frequently use discussion with the team as a means of arriving at a decision. Construction project managers will mainly operate along the top of the grid between the 'consensus manager' to the 'consensus autocrat' but may at certain times find themselves having to adopt styles located at other points in the grid.

The relationship between leadership and organisation culture has been referred to. The problem with this concept in relation to project organisations is that analysis of the organisational culture of construction project organisations does not appear yet to have been undertaken so its impact on leadership is not known. The concept of organisation culture in relation to temporary organisation structures rather than permanent evolving business organisations, is likely to need special consideration as no sooner will a culture begin to grow than it will be dissipated. Its effect on management style may be limited although the development of even a limited culture is far more likely to be affected by the management style of the project manager.

The varying requirements of leadership styles are readily apparent in project management organisations. For example:

- At the early stages of a project, the manager has to weld together the range of professional specialists involved in the project and lead them in balancing the conflicting objectives which will no doubt have emerged from the client organisation. Developing a viable solution which balances function, cost and time requires the project manager to create a climax in which everyone is free to speak up, make suggestions and criticise. To achieve this the project manager must adopt a human-relations orientated style of leadership. Once the proposed solution has been defined, the process of developing the solution into working documents becomes a much more structured process requiring a more task-orientated leadership style.
- During the process of developing working drawings, contract documentation, etc., which require a task-orientated approach, it may become apparent that a major problem occurs in translating the proposed solution. In which case, the project manager is unlikely to be able to solve the problem alone. The project manager will need to bring together professional specialists to discuss and resolve the problem and the project manager's style will switch from a task to human orientated style until the routine work is recommenced.
- The project manager's formal authority affects the style adopted. During the construction stage of a project under conventional contract conditions, the project manager will normally have to adopt a directive or autocratic style as the contractual context places specific obligations upon him or her. Such conditions make it difficult to adopt a human-relations style even when it appears to be the best way of approaching a problem. 'Partnering' is an approach which seeks to allow a more human-relations style which could be particularly effective in the construction stage.
- The degree of authority can also affect the project manager's style during the pre-contract stage. The authority which is held by the project manager who is an employee of the client organisation or who is given clear authority by the client, provides the project manager with a greater opportunity to be autocratic which is not the case if the project manager's authority is ambiguous. The former gives the project manager a greater range of leadership styles to use if he or she is capable of exercising such flexibility and if he or she has the ability to resist falling back on his or her formal authority in inappropriate situations.

Project managers' perceptions

It is really not possible to separate project managers' instinctive leadership styles from their view of the people they are leading. As a result the degree of flexibility of leadership styles of which a manager is capable may be inhibited by the fundamental assumptions made about the people being led.

The early work in this area was undertaken by Maslow (1954) who developed the famous 'hierarchy of people's needs' as illustrated in Fig. 10.4. This work defined what people were seeking from work and hence what motivates them and identified that:

● The needs hierarchy is based on needs, not wants.
● It operates on an ascending scale. As one set of needs becomes fulfilled, the next higher set comes into play.
● We can revert to a lower level. For instance, a person operating at level 4 or 5 will fall to level 2 if a feeling of insecurity takes over. Once the need is met, however, he will return to his former needs area.
● Needs not being met are demonstrated in behaviour. To create an environment in which motivation can take place, managers must therefore be able to recognise behaviour patterns in individuals and work groups. This means developing the ability to 'read' people and situations.

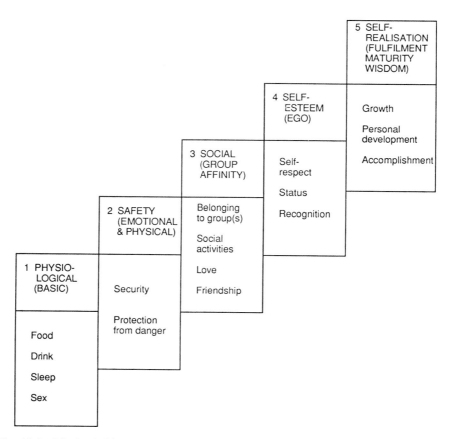

Fig. 10.4   Maslow's hierarchy of individual needs (Maslow 1954).

- To avoid apathy, which finally results when needs are unfulfilled, managers must be able to implement the right action at the right time.

This fundamental work was developed further by McGregor (1960) and resulted in his Theory X – Theory Y model referred to previously in Chapter 3. He discovered that managers' thinking and approaches were based on two different sets of assumptions about people. A Theory X manager instinctively adopts an autocratic leadership style because he assumes, amongst other negative features, that people are lazy and wish to avoid responsibility. A Theory Y manager instinctively adopts a democratic style as he assumes, amongst other positive features, that people are self motivated and wish to achieve and enjoy responsibility.

Herzberg's (1968) subsequent work on motivation reflected the same perceptions and he identified that by removing factors causing dissatisfaction a manager removes unhappiness but does not generate motivation whereas he was able to identify the factors which motivated people which correspond to Maslow's levels 4 and 5.

Each project manager will have a particular view of the people being led which will be somewhere in the Theory X – Theory Y scale and which will determine the project manager's habitual or instinctive leadership style. Recognition of this by project managers may allow them to be more flexible but if they are faced with situations in which they are uncertain they are likely to fall back upon their instinctive leadership style.

It is important to recognise that whilst the ideas of people's attitude to work are intellectually satisfying and difficult to disagree with at that level, they are difficult to apply consistently in practice. Eilon (1979) has pointed out that, if everyone is self-actualising then the task to be achieved is likely to be neglected when it is routine but vital to success.

Whilst the argument for leaders to use flexible leadership styles may be acceptable, there is an equally strong argument for those being led to adopt flexibility in their expectations from their work and from their leaders.

This work, whilst still useful, was the forerunner to the broader field of organisation behaviour. It should be pointed out that it is generally held that there is no such study as behaviour of an organisation but only a study of the behaviour of individuals in organisations (Naylor *et al.* 1980). The literature in the field draws heavily on industrial/organisational psychology and the social sciences, one of the more important aspects of which is motivation. Much of the focus is on the behaviour–performance–outcome relationship (e.g. Vroom 1964; Porter & Lawler 1968; Campbell *et al.* 1970; Naylor *et al.* 1980) which believes that the attractiveness of outcomes and individuals' expectation of success determine the amount of effort that they are willing to apply to achieve the goals. They also evaluate their own performances against the expected outcomes. Such themes are far more complex and sophisticated than those described earlier but it is not possible to develop them further here.

Leadership qualities

What then are the qualities which a project manager should possess to be a good leader? What are the qualities which allow the project manager to be sensitive to the different situations which arise on a project and be sufficiently flexible to use an appropriate leadership style?

These qualities can be split into characteristics and skills. The project manager's characteristics will in many cases determine how well he employs his skills. Examples of the characteristics which help to form good leaders in construction project management are:

- Integrity
- Preferred leadership style (tending towards democratic)
- Self-confidence
- Ability to delegate and trust others
- Ability to cope with stress
- Decisiveness
- Judgement
- Consistency and stability
- Personal motivation and dedications
- Determination
- Positive thinking
- Excellent health
- Openness and the ability to hear what others say
- Ease in social interactions with many types of people.

In terms of skills, the following are important:

- Persuasive ability
- Negotiation skills
- Commercial expertise
- 'Political' awareness
- Breadth of vision
- Integrative skills
- Ability to set clear objectives
- Communication skills
- Management of meetings
- Early warning antennae
- Skills of diplomacy
- The skills of discriminating important information.

The context of these qualities is the experience of the project manager. A broadly based experience is required of all phases of a project. The project manager will require an appreciation of all the specialist areas, whilst not needing to be a specialist in any. Nevertheless certain aspects will be more significant to the project manager such as contract strategies, cost and time control, money management including project finance and capital and revenue relationships. This type of experience can give project managers the confidence which a leader needs but equally importantly is the way in which it affects the team's perception of its leader. If project managers are seen to have the right experience, or more importantly, they have a successful record of achievement, be it by good leadership or good luck, the team will more readily accept them as leader. Project managers' acceptance in this way can account for much of their perceived leadership qualities.

It is generally accepted that there is no ideal leader for all situations. The variables involved in leadership behaviour and success are wide and unquantifiable. It is only possible to identify them and recognise the kinds of situations in which different combinations of attributes may be most beneficial. It is then necessary to require that leaders have the ability to vary the way in which they amalgamate their attributes depending on the situation in which they find themselves. Failing the leader's ability to be able to do that, it is necessary to change the leader when the situation changes. Neither alternative is really viable although some leaders do have some ability to change their style within limits. A transfer of leadership is for most organisations unrealistic which in any case could be extremely disruptive if allowed to happen frequently. The ability to select appropriate leaders in such situations must also be in question.

A respected leader has an amalgamation of characteristics, skills and experience which are recognised and respected by those he leads. Leadership is granted by the people being led and not by an organisational position. The effective leader influences not just subordinates but superiors and peers and in construction project management has the additional need to exercise such influence over people in other companies, which requires leadership characteristics of a very high order.

In trying to answer the question of how one spots a leader it is worth turning to Townsend (1984): 'They come in all ages, shapes and sizes and conditions. Some are poor administrators, some are not overly bright. One clue: since most people *per se* are mediocre, the true leader can be recognised because, somehow or other, his people consistently turn in superior performances.'

Project management agreements

Much writing and research on project management in general is undertaken without reference to formal agreements for the provision of project management services. This is because in many other industries project management is nearly always 'in-house' to the client organisation. The reverse applies in the development and construction sector and, as a result, formal agreements have to be drafted between clients and the firms carrying out project management. Such legal contracts provide the context within which project management services are provided and place constraints upon project managers and may inhibit the manner in which they manage the project. They may not act in the same way as they would if they were a member of the client organisation acting as project manager. For example, instead of adopting an appropriate leadership style for the situation, a project manager may fall back on the formal agreement with the client as the easiest way out of a difficult situation rather than seeking a creative answer. This 'soft option' would not be available if the project manager was 'in-house' to the client organisation.

Whilst formal legally binding agreements will not be entered into by 'in-house' project managers, they should have terms of reference which are similar in content. In fact the terms of reference given to 'in-house' project managers are likely to be more detailed than the contracts with external companies. External companies may seek to limit their responsibility whilst still wishing to appear to be carrying out all the functions of project management.

For example, the British Property Federation's (1983) (BPF) System for building design and construction identified the responsibilities of a client's representative who is the BPF equivalent of a project manager. The list, given in Table 10.3, is extremely detailed and reflects the content likely to be found in an 'in-house' project manager's terms of reference although it implies that it is

Table 10.3   BPF System. Schedule of responsibilities and duties of the client's representative.

Main responsibility	To manage the briefing, design, construction and commissioning processes on behalf of the client. Some or all of the responsibilities of the client's representative may be allocated to a member of the client's staff.
Limits of authority	The client's representative has no authority to vary the brief, the cost plan or the master programme except upon the instructions of the client.
Actions (1 to 6 are normally allocated to client's own staff)	1 Prepare initial viability studies 2 Investigate Development Land Tax liability 3 Negotiate with funding institutions, letting and publicity agents 4 Obtain legal advice and assistance on leasing arrangements, restrictive covenants and all other matters associated with ownership and letting. 5 Negotiate with ground landlord 6 Deal with future tenants' and tenants' queries and requirements 7 Prepare budget costs and present feasibility studies for the client 8 Prepare outline design brief 9 Conduct planning negotiations 10 Maintain records of meetings and all other activities 11 Amend brief as instructed by the client 12 Convene and chair project meetings 13 Check and approve consultant's and contractor's applications for payments

Table 10.3   (continued)

Actions (continued)	14 Amplify design brief for client approval
	15 Report options for sketch proposals to client
	16 Submit planning applications
	17 Arrange for client to place orders for soil investigations, structural and other surveys and models
	18 Sanction consultants' designs
	19 Prepare and maintain master programme
	20 Prepare and maintain cost plan
	21 Issue brief and amendments to design leader
	22 Obtain client authorisation for changes in master programme and cost plan
	23 Check professional indemnity insurance of consultants
	24 Discuss buildability and technical design with specialist contractors
	25 Approve proposed named subcontractors
	26 Produce monthly reports to client giving forecast final costs and forecast completion date
	27 Arrange appointment of adjudicators and bring in their services where necessary
	28 Authorise amendment of priced design programmes and activity schedules
	29 Approve preferred components
	30 Submit final design proposals and revised cost plan
	31 Draw up list of qualified tenderers
	32 Send out tender documents
	33 Receive tenders and obtain clarification as required
	34 Prepare tender award recommendation to client
	35 Assemble contract documents, raise bonds and obtain client's and contractor's signatures to building agreement
	36 Arrange orders for long delivery components
	37 Obtain contractor's insurance certificate
	38 Prepare valuations and prepare final account
	39 Issue variation orders and instructions
	40 Obtain client approval to costs of variations and claims when limit of authority exceeded
	41 Deal with insurance claims
	42 Receive occupation certificates and clearance from health and safety and fire officers
	43 Issue certificates of extensions of time
	44 Issue taking over and final certificates
	45 Ensure that defects are remedied
	46 Act on early warning notices
	47 Prepare client's commissioning and equipping programme and obtain maintenance manuals and as-built drawings
Advice to client	48 Advise on selection of design leader
	49 Advise on need for and selection of other consultants
	50 Advise on terms of appointments of consultants and design leader
	51 Check and approve priced design programmes of consultants

Table 10.3   (continued)

Advice to client (continued)	52 Advise on work, if any, to be carried out by direct contract with the client
	53 Advise on appointment of supervisor
Decisions and instructions	54 Instruct consultants on research, surveys and feasibility studies
	55 Decide how the design is to be split between consultant and contractor
	56 Decide type of information to be produced by consultants
	57 Instruct consultants on reporting and recording procedures
	58 Decide who will be the adjudicator
Monitoring	59 Monitor the performance by the design leader, consultants, supervisor and contractor in their performance of duties and discharge of responsibilities as set out in their respective schedules of responsibilities and duties
Liaison and co-ordination	60 Liaise with lawyers on critical restrictions and leasing arrangements
	61 Co-ordinate tenants' queries

appropriate to either. It reflects the functions of project management in Chapter 8 but breaks the functions down into a much finer level of detail. Such a level of detail may be inappropriate to a formal agreement as it could create too many opportunities for claiming that certain work is not included and should either be done by some other consultant or be subject to additional payment. This could generate inappropriate attitudes on the part of the client, project manager and consultants. Nevertheless, whilst not having widespread use, it provides a very useful checklist for the preparation of terms of reference of formal agreements although it does of course reflect the particular aspects of the BPF System, e.g. references to design leader and named subcontractors.

As mentioned earlier formal agreements with external independent project management companies will not normally be so detailed. They will usually consist of two parts – the Memorandum of Agreement and the Conditions of Engagement.

The Memorandum of Agreement is the formally drafted agreement that the client has appointed the project management firm and the firm has agreed to accept the appointment. It will describe the project, for example what and where it is, the estimated cost, which types of consultants are to be engaged and that a contractor will be appointed.

Important aspects will be the duration of the project manager's appointment, the relationship of the Conditions of Engagement and relationship with other consultants. Such a clause could be:

'The client hereby appoints the partnership for a period of *n* months from the date first above written subject to and in accordance with the Conditions of Engagement attached hereto and the partnership agrees to discharge its rights duties obligations and responsibilities in accordance with this agreement and the said Conditions of Engagement with due regard to the rights duties and obligations of the design team and the contractor employed for the project.'

Clauses should be included on arrangements for termination of the agreement as should a clause stating the fee or how it should be calculated together with any arrangements for its adjustment and any provision for extending the agreement beyond the expiry date and how payment for any extension should be calculated. The agreement would be signed and witnessed.

The Conditions of Engagement would normally include the following:

(a) The responsibilities of the project manager
(b) The parameters within which the project manager should operate
(c) The services to be provided
(d) Exclusions – the services not provided
(e) The formal reporting requirements
(f) Provision of facilities
(g) Compatibility of agreements
(h) Arrangements for fee payment
(i) Names of the client's contact and project manager.

(a) The responsibilities

This is perhaps the most difficult section to draft as it can vary from the very simple to the very detailed, neither of which is likely to be satisfactory to either party. Should there be a dispute which leads to litigation it is likely to be these clauses which are subject to interpretation by the courts.

The following is an example which illustrates the problem of drafting.

'The partnership shall be responsible on behalf of the client for:

(i) the management of the design, construction, equipping and commissioning of the project by organising the design team and the contractor (as led by the architect) and co-ordinating those departments of the client directly concerned in the project;
(ii) using their reasonable endeavours to secure completion of the project within the time and cost limits determined by the client from time to time;
(iii) the maintenance of communications with and where appropriate the co-ordination of the parties and organisations directly involved with the project.'

The use of such words as 'management', 'organise' and 'co-ordination' is practically unavoidable yet they are extremely difficult terms to define with any precision. Similarly 'reasonable endeavours' and 'communications' fall into the same categories. There has therefore to be a high level of trust and goodwill between the client and project management firm as a recourse to litigation should a dispute occur is likely to be unsatisfactory to anyone other than the lawyers.

(b) The parameters

Constraints on the manner in which both the client and project manager can act are likely to be needed. They will include such aspects as the need to work in conformity with policies and procedures laid down by the client which will include the client's financial and contracting systems and will require the client to inform the project manager of them. In addition, such parameters will also require the project manager to act in accordance with the client's grievance procedures in supervising client's staff and would probably also place the responsibility for appointing, dismissing and paying consultants with the client when the consultants are from practices separate from the project manager.

(c) The services

The extent to which these are detailed will vary between agreements as a result of negotiation between the client and project manager but would normally cover the functions of project management listed in Chapter 8. This section is usually characterised by such terms as 'monitoring', 'co-ordinating', 'liaising' and 'advising', all of which are difficult to define and hence difficult to determine whether they have been executed effectively.

(d) The exclusions

The exclusions clarify what is excluded from the agreement as it is frequently easier to be more precise with exclusions than with inclusions. Examples of the aspects which may be excluded from the project manager's role under the agreement are:

(i) Specifying and procuring labour, material and equipment for the project.
(ii) Legal advice, participation in arbitration or legal proceedings.
(iii) Any of the professional functions, e.g. design, surveying.
(iv) Approval of workmanship in construction.
(v) Management of the client's staff and associated industrial relations.
(vi) Securing funds for the project.
(vii) Auditing.
(viii) Negotiating statutory approvals.
(ix) Public relations in connection with the project.

These only indicate areas which may be excluded. They could easily be incorporated should both sides wish. Similarly the degree to which they are included or excluded can be negotiated. For example securing funds is excluded but normally the preparation of applications for funds would be included. Similarly assistance in preparing publicity material may be included.

(e) Reporting

It would normally be important to both parties to have the reporting procedures clearly laid down. This would state the frequency of reports, often monthly, and their content.

Content would normally cover such aspects as:

(i) An overall review of progress relative to key dates established in the programme and the budget.
(ii) Explanations of deviation from the programme and budget.
(iii) A statement of the decisions and actions required of the client in the near future together with a review and recommendation from the project manager.
(iv) A review of the more important problems and potential problems perceived by the project manager with recommendations where appropriate.
(v) A prediction of progress during the next month.

(f) Facilities
If facilities for the project manager are required to be provided by the client details should be included. For example the provisions of office accommodation, telephone, fax etc.

(g) Compatibility
It is beneficial to include a clause requiring the client to ensure that his agreements with any other parties which impinge on the project management agreement are compatible with it.

(h) Fees
The amount of fee will have been included in the Memorandum of Agreement but the details of how it is to be paid, e.g. by stages, and arrangements for any increase to offset inflation should be included in the conditions.

(i) Names
It may be appropriate to name a member of the client body and specify that the named person shall act with the full authority of the client. Similarly it may be appropriate to name the project manager, particularly when the project management firm has been awarded the project on the basis of a presentation which included the proposal of a named individual as project manager. Nevertheless, however attractive this idea may be it could be impracticable on long duration projects and even on shorter projects it can create difficulties if one of the named persons leaves the firm. For similar reasons, if a specific project manager is named it would be sensible for the conditions of engagement to provide for certain of the duties to be delegated. It has been suggested that, in the absence of an express entitlement to delegate some of his or her duties, the project manager in attempting to delegate could be in breach of contract (Knowles 1986).

This outline only indicates the areas which may need covering in project management agreements which will need to be negotiated separately for each project to reflect the particular needs of the client and the project. They form an important contract and should be formally drafted by a lawyer although model Memorandum of Agreement and Conditions of Engagement and Guidance Notes for their use are published by the Royal Institute of Chartered Surveyors. Whilst it is always expected that there will be no recourse to the small print and litigation, nevertheless the rights and obligations of the parties should be spelled out.

It should be pointed out that if project management is provided by a multi-disciplinary practice which also provides some or all of the design consultancy services or even construction then the form of agreement is likely to be considerably different from the outline given above.

The legal context

Legal rights and duties will arise from the appointment of a project manager, in particular the law of contract and of tort govern the manner in which the project manager carries out his or her duties. Roger Knowles (1986) reviewed the position in the UK and found that due to the comparative newness of the position of the project manager there was little case law with which to gauge liability. Knowles felt that there was still a need to define the exact role of the project manager in order for it to be incorporated into agreements but he also recognised that there was an argument for agreements to include only objectives leaving project managers to decide what they must do to achieve them.

Two particular duties which may be implied in agreements and which could cause problems are responsibility for design and responsibility for quality of workmanship. Whilst most project managers would anxious to assert that they have no responsibility if things go wrong with the design or the construction of the work, there may well be liability. For instance the BPF System Conditions of Engagement require the client's representative to manage the design and 'to study the drawings and specification in order to sanction them' and 'to examine material put forward by the design team leader to ensure that it meets the brief'. In all probability the project manager would be expected to notice any obvious design faults and if he or she failed to do so may have a liability. Case law exists for similar circumstances but involving persons other than project managers. Knowles suggests that if project managers enter into their agreement with the client without too much thought they may find they have undertaken an obligation to project manage a scheme to ensure the building is completed to time, to specification and to budget. Such an obligation may be deemed by a court as guaranteeing the result, and any failure may amount to a breach of contract.

The tort of negligence is concerned with breach of duty to take care in which the plaintiff was owed a duty of care and has suffered damage as a result of that breach. Knowles examined to whom the liability of the project manager might extend. He found that again there was no directly relevant case law but pointed out that there certainly existed a duty of care to the employer (client) under both the conditions of engagement and the law of torts. It will be for the courts to decide who else is owed such a duty. However in looking for clues in associated case law he suggested that the main contractor may be in this category but felt that subcontractors may be too remote. He also pointed out that liability covers economic loss in addition to the accepted losses of injury to health or damage to property. Knowles makes no reference to the project manager's liability to the other consultants but as he concludes 'liability under the law of torts will be a matter of decision by the courts'. No doubt it is this uncertainty which creates the situation that when a consultant completes a form for professional indemnity insurance he finds that project management is subject to a separate declaration, agreement and premium with many insurers.

Chapter 11
Organisation Structures

Introduction

The conventional approach to project organisation with the architect as team leader, responsible for both design and management of the project with the contractor appointed on the basis of a competitive tender, still predominates in the UK and many other parts of the world (Mohsini & Davidson 1992). However, in recent years variations on this method have emerged, as well as some rather more innovative approaches to project organisation.

This chapter examines the contribution of alternative approaches to the solution of organisational problems against the features of project organisation previously identified. These features were:

(a) The relationship of the project team to the client organisation and the client's influence upon the decision points.
(b) The degree of interdependency of tasks and people generated by the project and the organisation structure.
(c) The degree of differentiation present within the operating system (which ideally should be reduced to a minimum). The level to which it can be reduced will be constrained by the nature of the project.
(d) The level of integration provided by the managing system and the complexity of the managing system itself. Over-elaboration can lead to severe differentiation within the managing system, which should have the capability to match its integrative effort to the degree of differentiation present in the project.

In practice there are three major components to the organisation structure of projects:

(a) The client/project team integrative mechanism.
(b) The organisation of the design team.
(c) The integration of the construction team into the process.

A number of options are available within each of these categories and this results in a large number of possible combinations. The whole range cannot be considered here but the more likely alternatives are analysed. Of course some options within (b) would not be used with some options within (c), e.g. a conventionally organised design team with a design-and-build contract. However, many others are combinable, and have an effect one upon the other, e.g. a conventionally organised design team and management contracting, which could have a profound effect upon how the design team is organised.

Client/project team integration

This aspect has been discussed earlier, and, as was pointed out, the variety of organisation structures of client organisations is vast. It will not be possible for the project team to affect the client's organisation structure other than marginally. It will be up to them to organise themselves to fit in with the client's organisation. It is therefore a case of the project team organising itself so that it has the capability to understand the client's firm and its environment in order that it can respond to the client's requirements and any changes that may be dictated by the client's environment during design and construction.

It will be easier to integrate with some clients than with others. Where the client has in-house expertise in construction, it is to be expected that the dovetailing of the project team with the client will be easier. On the other hand, integration with a client who does not have in-house expertise or, even more difficult, one who has not built previously, will be more of a problem. The response therefore has to come from the project team and the structure of the design and construction teams should be set up so as to reflect the difficulty of integration with the client. For example, if the client is experienced in construction and has in-house expertise, the client may appoint a project manager with experience of the construction industry from the client's own staff. In such a situation, given the right qualities in the project manager, it may well be that a conventionally structured design team under the direction of the project manager would be appropriate and economical. There could well be no case for the appointment of a further project manager from the design team.

Alternatively, the client may be building for the first time and embarking upon a complex project, e.g. the rebuilding of a processing plant. Because of the naivety of the client and the complexity of the project, it may be advisable to appoint a project manager in an executive capacity from outside the client's organisation.

Similarly, as referred to previously, the authority of the project manager will vary depending upon the attitude of the client to delegation. The extent of delegation of authority is likely to be strongly influenced by whether the project manager is 'in-house' to the client's organisation or external to it. There will be a tendency for clients to delegate more to an 'in-house' project manager and such a project manager will also have greater access to the internal workings of the client organisation. Even in such a case delegation may not be high if the client's organisation is hierarchical and bureaucratic.

There can be no hard and fast rules for the integration of the client and the project team. The mechanism that is selected should be the result of an analysis of the client's organisational structure, the client's needs and the complexity of the project. The objective of such a mechanism should be simplicity within the constraints of the need to identify clearly decision points and the client's involvement with the decision-making process.

Design team organisation

Conventional structure

The conventional structure of architect responsible for design and management with other consultants acting for the architect and with estate management functions being directly responsible to the client is illustrated diagrammatically in Fig. 11.1. In such an arrangement the contractor is normally appointed after the design is substantially complete, usually by competition, although the contractor may be appointed on the basis of a negotiated tender or by some other means.

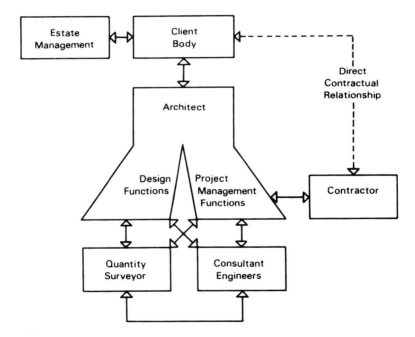

Fig. 11.1   Conventional structure.

In many cases each contributor will be from an independent professional practice, the contractor also being independent of the other contributors, yet the contributors will be interdependent in terms of the project. The more complex the client organisation and/or the project, the more interdependent will be the tasks to be carried out in achieving the project and the more the contributors will rely upon each other to carry out their work satisfactorily.

Such a structure produces a high level of differentiation between the contributors, which demands a high level of integration. The problem of providing the appropriate level of integration is compounded by the fact that the managing system is not differentiated from the operating system. That is, the architect is attempting to fulfil dual roles. One is in the operating system – design – the other in the management of the project. There is therefore a high potential for someone in this position not to be able to exercise objectivity in decision making. In addition, whoever is in this position is placed under severe pressure by being required to undertake tasks that frequently require what are often incompatible skills – design and management. This does not mean that adequate project management cannot in any circumstances take place in such a structure, only that it may be extremely difficult to achieve and will require a person of exceptional talent to be able to fulfil both roles satisfactorily in the complex environment within which construction takes place.

Although this does not have to be the case, such a structure has a tendency to restrict access of the other contributors to the client and hence the decision-making process. The perceived personal relationship between the client and the architect, particularly with clients new to construction, can inhibit the client from approaching the other contributors for direct advice. As there is no one solely in a project management role, there is a danger that apposite advice is not taken, which will be to the detriment of the outcome of the project.

Integration within the design team can therefore be difficult to achieve in this structure, as can the integration of the design team with the client. This situation is made even more difficult by appointing the contractor in competition. The direct contract with the client which this produces reinforces the contractor's differentiation from the design team. Similarly, the frequent exclusion from the design team of the estate manager also adds to differentiation. Whilst the conventional system can be seen to be problematic, it has been argued that this traditional approach may still be just as effective as more novel approaches (Bresnen 1991) and that such approaches are often re-selected in preference to the uncertainty and disturbance that may ensue with a departure from normal practice (Bresnen & Haslam 1991). The latter is, of course, only the case when the client is experienced with the system.

The situation can arise in which the quantity surveyor, estate manager or engineer is the first contributor to be appointed and subsequently has advised the client on the appointment of the other consultants. In each of these cases the result will tend to be similar to the situations described in Fig. 11.1. That is, the first appointed contributor assumes project management responsibilities alongside the professional functions, leading to a potential lack of objectivity in weighing factors from other contributors and to integration problems equally as difficult as those described above.

The degree of differentiation would be reduced if all the design team contributors were from the same interdisciplinary practice but, even within such a practice, if its members are organised on conventional lines with the team leader exercising both professional and project management functions, the main hindrance to objectivity and integration would still remain. However, such a practice has a better opportunity of overcoming problems created by differentiation and of generating sound integration for individual projects than if projects are designed using independent practices. A parallel situation would exist if all the design contributors were in-house to the client's organisation (e.g. a government department, a local authority or a large industrial concern). A major additional advantage in this situation would be the potentially high level of integration with the client, as client and design team would be under the same organisational umbrella.

Taking this argument a stage further, the organisation that should, theoretically, have the least differentiation and the greatest opportunity to achieve full client integration is one that has a construction capacity as well as a design capability within the client organisation, for example a local authority direct labour organisation or a developer/contractor. However, problems of control and motivation have been shown to exist in the case of direct labour departments whereas developer/contractor organisations have been successful in many cases.

Non-executive project management

A structure often employed by interdisciplinary practices, either private practices or in-house to the client's organisation, is one that includes a non-executive project manager (Walker 1976), sometimes called a co-ordinator, who operates in parallel with the other contributors, as illustrated in Fig. 11.2. The role undertaken by the person in this position is based upon communication and co-ordination activities and is not concerned with decision making. In these circumstances responsibility for the success or failure of the project will be with the firm or the particular in-house department and not with the non-executive project manager within the team as would be the case with a project manager acting in an executive capacity. There is therefore less pressure for the project manager or the firm to define the project manager's role and authority. What pressures there are will be internal to the firm or departments, depending upon how they see the role of the project manager within the team.

Such a role is unlikely to have a significant effect upon the quality of integration of the design team with the client's organisation but could, if exercised with skill and received positively by members of the design team, assist in integrating the design team. If exercised unskilfully or in an unco-operative climate, it could emphasise differentiation within the team without contributing to integration. The authority of the non-executive project manager is likely to be weak and hence his or her ability to contribute will be determined by the commitment of the firm, his or her informal authority, and the attitude of the individual members of the design team to his or her role.

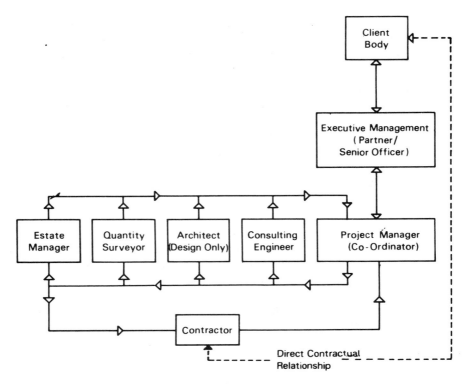

Fig. 11.2   Non-executive project management structure.

A non-executive project manager could be used where the contributors are from separate practices. Such a role is likely to be undertaken by one of the independent firms contributing to the design, although it is conceivable that it could be undertaken by a firm solely devoted to project management. The result is unlikely to contribute much to the project management process, although any improvement in co-ordination and communication would be of benefit. The lack of an executive function would mean that project management would be split between the non-executive role and the decision-making role, which would still be undertaken by the architect, partner of the lead consultancy or senior officer in the case of a public authority. The potential benefit of improved co-ordination and communication may well be more than offset by the complexity of the management system which emerged. The major management role will still not be separated from the operating system and, in fact, if the non-executive role were shared with an operating role, this situation would be further compounded.

Executive project management

An executive project management role (Walker 1976) is undertaken by a firm or person independent of the other contributors to the process, as illustrated in Fig.

11.3. Similarly, if the design team is part of an interdisciplinary practice or in-house to the client, it will be undertaken by a member who is also independent of the other contributors. In such a structure the project management activity occupies a dominant role in relation to the other contributors, and although they operate as a team, the project manager will make the decisions that are within the purview of the contributors. He or she will tend to be the sole *formal* point of reference to the client for the purpose of agreeing and transmitting the decisions that must be made by the client. In addition, the project manager will be concerned with controlling, monitoring and maintaining the project team, as discussed previously. These activities are far more dynamic and purposeful than the co-ordination and communication activities of the non-executive project manager and do, of course, subsume them.

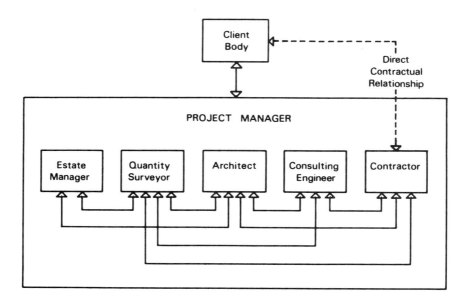

Fig. 11.3   Executive project management structure.

It is necessary that the firm or person undertaking this role ensures that responsibility and authority are clearly established with both the client and the other contributors to the project. Although this is difficult to achieve, the benefits of producing a situation in which the roles of the contributors are clearly established are significant. It is, of course, unrealistic for the project manager to accept responsibility for the technical work of the specialist contributors, but responsibility for progress and for budget control are possible given the authority to report back to the client if contributors are not performing satisfactorily. However, in practice, the project manager should work in a collaborative manner with the contributors and a major role would be one of facilitating the work of all the contributors so that the project is developed by a team approach. The project manager's primary concern would be that appropriate decisions are taken by both the client and the project team at the right time.

In an interdisciplinary practice or in-house to the client, the responsibility for the project is clearly with the firm or the department, respectively. The authority of the project manager will be decided internally in the case of an interdisciplinary practice and it is unlikely that the client would exert as great an influence as it would over an in-house department.

The major benefit claimed for the executive project management structure is that management becomes clearly separated from the operating system. That is, no person is charged with carrying out both design activities, be they architecture, engineering or quantity surveying, and project management activities. This allows concentration upon the management needs of the project and makes it possible for conflicting professional advice to be considered more objectively so that decisions which are in the best interests of the project as a whole can be made or recommended.

The structure facilitates integration with the client because the person responsible for managing the project within the client's organisation can readily identify the management responsibility within the design team and is likely to have empathy with the person in this position. This should facilitate the decision-making process, particularly within the client's organisation. A major role of the project manager would be the planning and programming of the project, which would include identifying the contributors and their roles. In doing this the project manager should recognise the differentiation generated by the particular project. The project manager should be in a position to judge the integrative demands and should design mechanisms to cope with them. It is to be expected, therefore, that the project manager will be intimately involved with the client in determining how the organisation for the project should be structured and in deciding the firms and people who will undertake the various roles.

In certain circumstances the client's organisation has personnel who can undertake the executive project manager's role without the need to appoint someone from outside the client's organisation. This arrangement can contribute significantly to the ease of integration between the client and the design team, provided, of course, that the project manager has sufficient status within the client's organisation and can command the respect of the design team. If not, there may be a tendency for the design team to bypass the project manager and seek higher authority in the client's organisation. In this case the role of the project manager would be seriously undermined with resultant confusion in the decision-making process. A similar situation can arise when the project manager is outside the client's organisation, and the solution will be in the attitude which the client takes to resolving the situation.

The person appointed from within the client's organisation to liaise with the design team when an executive project manager has been appointed from an outside organisation is also often called a project manager and it is therefore important to recognise their different roles.

It is also important to recognise that if a person in the design team who is a member of one of the firms contributing in a professional capacity (e.g. quantity surveyor, engineer) is given the title of project manager and ascribed specific responsibilities and authority in this capacity, this does not constitute an executive project management structure as described here. It is in fact a variation on the conventional structure as they will be acting in the same way as the architect traditionally acts, and some of the objectivity of the executive structure would be lost.

In the early 1980s the British Property Federation (1983) devised a system, which was not dissimilar from the executive project management approach, in response to their perceived need to improve the conventional structure as it had become increasingly concerned about 'problems of poor design, inadequate supervision, insufficient choice of material and contractual methods which caused delay and increased cost.'

It restructured the traditional organisation and hence relationships between the participants to projects by introducing a client representative and a design leader. The client's representative was akin to a project manager and 'assists the client to develop the concept of the project; he advises on the selection of consultants, he resolves conflicting priorities, instructs the consultants and contractors and safeguards the client's financial interests'. The client's representative may be from within the client organisation or an appointed individual or professional firm from outside the client organisation.

Design consultants were responsible for all aspects of design, e.g. function, structures, services, cost, but they were co-ordinated by a design leader who had overall responsibility for all aspects of pre-tender design and specification. Although the system laid down the individual responsibilities of the client representative and the design leader, the duties were transferable between these managers. The client representative was responsible for all aspects of management during construction but worked in conjunction with the design leader and a supervisor. The supervisor, who was appointed by the client, monitored the work to ensure the building was constructed as designed and specified in the contract documents.

It was claimed that the BPF system was innovative and challenging to the professions and industry. In terms of its organisational structure its innovation lay in the introduction of additional levels of management. This development subscribed to the idea of an open system moving towards a higher level of organisation with greater differentiation and a corresponding need for greater integration. The system also had the attribute of flexibility so that it could be tailored to suit the particular project to which it was applied.

In fact, the most dramatic innovations of the system were in its procedures rather than in its organisation structure. Many of the features of the organisation structure had been recognised for some time. The novelty lies in the manner in which they are amalgamated into the system and integrated with the more innovative procedural initiatives such as an adjudicator to settle disputes, fixed fees plus incentive fees for consultants, elimination of the bill of quantities, contractor designs, the use of prices linked to construction activities and the elimination of nominated subcontractors. Whilst not in wide use, the system's ideas had a strong impact and influenced the development of project management generally.

Integration of the construction team

The degree to which the construction team can be integrated into the process at the design stage is determined by the tendering arrangements which are made for obtaining the price for construction. This is obviously a key decision as it has a fundamental effect upon the organisation of the whole process. The most common method is by a lump sum competitive tender after the design has been substantially completed. A number of other methods are available, but still not commonly used, which provide for more constructive integration of the contractor.

Competitive lump sum tendering after completion of the design provides the least opportunity for integration as this method requires that the contractor who is to construct the project cannot be involved in the design stage. In addition, during construction it is often difficult to integrate the design team with the construction team as the split between design and construction appears, in many cases, to create a psychological barrier between the two groups. The contractor will often feel that the design has been carried out by people who do not understand construction methods and who seem to be providing the wrong drawings at the wrong time. The designers may adopt the view that the contractor is only concerned with profit and not with providing the service that will provide the project they require. Whether such views are correctly held or not does not overcome the fact that on conventionally organised projects, the greatest degree of differentiation occurs between the designers, normally represented by the architect, and the contractor.

If the project is managed by an executive project manager, then the greatest integrative effort is likely to be centred around this interface. If the project is organised conventionally, problems of integration at this point will be extremely difficult to resolve if the architect is both designer and project manager and has to also integrate with the contractor. The difficulties of integrating this interface are compounded still further by the use of subcontractors, both nominated, named and domestic. Subcontractors nominated or named by the architect will have a strong allegiance to the architect while having to work in a contractual arrangement and under the direction of the contractor. Domestic subcontractors hired directly by the contractor will tend to hold the same views as the contractor towards the designers.

Many of the recently developed methods of appointing contractors have been aimed at allowing them to be better integrated into the design team while still allowing an element of competition in obtaining a price for the project.

Two-stage tendering

In order to maintain competition in a similar form to the conventional method, yet allow the contractor to be involved to some degree in the design stage, two-stage tendering emerged in the late 1960s but has not developed to any great extent. In the first stage, selected contractors are invited to tender. Their tenders are based on a notional bill of quantities in which the items are fully described and the quantities are hypothetical but of the order of the magnitude anticipated in the proposed project. The successful tenderer is then involved in the further development of the design as a member of the team.

A bill of quantities is prepared for the fully developed scheme and is priced by the successful tenderer using the rates, where applicable, in the first-stage tender and negotiating other rates on the basis of the original tender. The result of this process is the price for the project.

This approach assists in the integration of the contractor into at least part of the design process but does not fully exploit the potential benefit to the project of involving the contractor from the beginning. Perhaps one of the greatest benefits is the opportunity to involve the contractor in selection of the specialist sub-contractors.

Designers are often sceptical about the contribution a contractor may make to the design of a project but this is one of the aspects the project manager must overcome in integrating the contractor into the design stage. The advice the contractor can give regarding the constructional implications of design decisions and construction methods and processes is likely to be recognised by the designers only after the event. This makes the project manager's task that much more difficult. It also illustrates the fact that integration of the contractor into the design process is less likely to happen in the conventional process where the decision to integrate would have to be made by designers.

Two-stage tendering represents a 'trade-off' between integration of the contractor into the design stage against a conventional approach to competition. It was one of the earlier approaches to integration of the contractor.

Serial tendering

Serial tendering is used to obtain tenders for a number of similar projects. Contractors bid on the basis of a notional bill of quantities and normally the lowest is accepted. The prices in the notional bill are used for a series of projects, the number, timing and size of which are indicated to the tenderers before bidding.

The actual price for each separate contract is calculated by using the rates submitted in the notional bill. On the face of it, this procedure allows the contractor to be integrated from the beginning of the design of each building in the series for which the contractor has been successful. However, this is rather artificial as the majority of the design decisions will have been made before production of the notional bill so that although the contractor may be integrated, the contractor's effective influence on the design stage is limited. Nevertheless, there is a distinct advantage over the conventional approach as it allows discussion with the contractor about such things as subcontractors, plant, programming, etc. during the design of each project in the series.

Negotiated tenders

The use of negotiated tenders does not rely upon a competitive element in selecting the contractor. The contractor is selected on the basis of reputation and will probably have worked satisfactorily for the client and/or design team previously. The price for construction will be agreed with the contractor following negotiation between the quantity surveyor and the contractor. There are a number of variations in the approach adopted, often including some types of target cost. Whichever approach is used, the effect on the organisation structure for the project will be similar. Under this arrangement the contractor can be involved in the design process at whatever point the client or project manager decides.

Integration of the contractor can achieve the highest possible level using this approach, but on many negotiated contracts the contractor is still not brought into the process until the design is well advanced and one of the major advantages of this approach is lost.

Naturally, such an arrangement requires a high level of trust between the client, design team and construction team. It is often said that a negotiated price will be higher than a competitive one so the integration of the contractor may be traded off against a higher price. However, the benefits to be gained from integration – earlier start on site and earlier completion, constructionally sound designs, cost-saving advice from the contractor, etc. – may more than counterbalance the lack of competition. Within this framework, subcontracts may be negotiated or competitive, giving a facility for closer integration of subcontractors if their use is considered beneficial.

Management contracting

This approach is fundamentally different from those previously described, as it is a method of integrating contracting expertise into the design stage of a project on a fee basis rather than a method of obtaining tenders and letting contracts. The objective is to incorporate the management contractor into the design team on an equivalent basis to all the other consultants. The management contractor would be responsible for the construction work, all of which is carried out by sub-contractors who may be appointed in competition or by negotiation.

The project is split into 'packages' for tendering purposes. Design, cost control and construction are integrated into the overall programme. The contractor provides advice on the availability and procurement of materials and components and the 'buildability' of the proposed design. The management contractor is involved in the compilation of the cost plan for the project and monitors and provides financial data concerning the project during the construction phase. The management contractor does not carry out the construction work directly but may provide certain central facilities (e.g. canteen, welfare, scaffolding). However, the management contractor sees its role as that of a manager.

It can be seen that management contracting is a positive approach to the integration of construction expertise into the design process. Its main trust is management of the construction aspects of the project in both the design and construction phases with status equivalent to that of all other professional contributors.

Whereas the subcontractors will be in a strongly differentiated position similar to that of the contractor in the conventional process, the management contracting activity will be primarily concerned with integrating them. This is rather different from the situation with a conventional structure as the management contractor's objective is that of achieving satisfaction for the client and does not have the entrepreneurial interest of the contractor employed under a conventional arrangement. This is in fact transferred to the subcontractors but in this case there is an integrating mechanism acting directly for the client.

The person responsible for project management will have the task of integrating the management contracting activity into the design team. If this is the architect, then it may be difficult, although it should not be so if the architect instigated the use of management contracting in the first instance. However, frequently it is the client who decides that management contracting should be used. In this case it is important that the remainder of the team is structured in a compatible manner. The allegiances and attitudes at large in the construction industry may make extremely difficult the integration of management contracting into a design team whose other members are unsympathetic to the idea. This may mean that either a strong client or a project manager with sound authority is necessary to gain greatest benefit from this approach.

Separate trades contracting

Separate trades contracting is used as a generic title for approaches that can be seen to be similar to the idea of management contracting. In each case there is no main contractor appointed for the project but instead a member of the client's organisation or fee-earning members of the project team organise subcontractors to undertake the work. The idea is not new: Scotland was the last region of the UK that used it as the normal method of arranging the construction stage and, although used only rarely and in specific circumstances in the UK now, it is still regularly used in other parts of the world. A particular form in the UK was documented by Thompson (1978), referred to as *alternative methods of management* (AMM), in which the site architect/manager was responsible for running the site and directed the activities of subcontractors either through their supervisor or directly to the men on site. Although there was no main contractor, there may be a need for a general builder to work alongside specialist contractors and to provide some central services. The subcontractors were appointed by competition or negotiation for packages of work. In effect the site architect/manager replaced the main contractor's site agent and provided the site with direct and constant design supervision. The site architect was often supported by a quantity surveyor who arranged the contracts and must have ready access to the client whose involvement on site was higher than on conventional projects. It was claimed that the main advantages were that communication was as direct as possible from client to architect to tradesman, that the human element was all important and the client's interest was best served by people committed primarily to the client's project rather than their profession or trade. Compared with management contracting, AMM does not incorporate construction expertise as such in the design stage but relies upon the ability of the architect in this respect.

Other separate trades contracting approaches are 'cost plus contracts' of various types. They are basically similar to the above, except that subcontractors are reimbursed their actual expenditure plus profit, often only within set total cost targets. The management implications are similar to those outlined above and demand similar levels of involvement of designers in managing the construction stage. They have not received the detailed attention in recent years that management contracting has, as they are really contractual arrangements and not approaches to management. They are rarely used except for emergency works as they can be expensive and can, of course, also be used for general contracting as well as for separate trades contracting.

Design-and-build

Design-and-build contracts are arrangements that do not separate design and construction as one firm offers the total package of design and construction.

Akintoye (1994) identifies six variations of design-and-build. Traditional design-and-build is defined as arrangements in which the contractor accepts total responsibility for both the design and construction of the project. Other types have varying degrees of involvement of the contractor with management of the design process and involvement in the actual construction of the project, including package deals and turnkey contracts which can be even more all-encompassing than traditional design-and-build. Worthy of particular note is 'novation design-and-build' in which the client's consultants, who have developed the project to the point of appointment of the design-and-build contractor, are passed to the contractor for the completion of the project. The opportunity to provide effective integration of the process is theoretically higher in design-and-build approaches than in more conventional methods although using the novation method presents less opportunity than traditional design-and-build.

There is a longer experience with design-and-build for civil engineering projects and many contractors which specialise in civil engineering work have strong in-house civil engineering design capability. There is also a long history of civil engineering projects giving the contractor an opportunity to submit an alternative design for all or part of the project to allow contractors to increase the competitiveness of their bids. The discussion which follows is couched in terms of building projects only, for simplicity, but the points made apply equally to civil engineering projects.

In the UK the large majority of firms offering this type of service originated as contractors and many also offer competitive contracting as well as a design-and-build service. There is therefore a tendency for firms to be orientated towards construction activity, which may have detrimental consequences for the integration of design and a subsequent effect upon its quality. The relationships that emerge may be as shown in Fig. 11.4. For such a structure the client would need to be assured that the construction management emphasis is not allowed to dominate the project management needs of the project. A structure which is more likely to be acceptable to the client is shown in Fig. 11.5 in which it can be seen that project management is allowed to dominate, and design and construction management are integrated in an equivalent relationship.

This latter arrangement is only likely to be adopted by a construction firm which has an 'in-house' design capability that is sufficient for the project. If the firm has to 'subcontract' design, then the relationships that emerge are likely to be similar to a conventional arrangement in terms of the difficulty of integration. Contractors generally prefer to subcontract design (Walker 1995) due to the greater access to a wider range of design skills provided by this approach and the reduction in risk associated with not having designers on their payroll. However, the client will still retain the advantage of having only one company responsible contractually for the whole project. Similarly, the responsibility for sub-contractors for specialist construction will be totally with the design-and-build firm and integration will not normally be complicated by nominated sub-contractor relationships.

Design-and-build may potentially provide the most effective integration but there remains difficulty in integrating the project team and the client. The client needs to protect its position so that the project it receives on completion fulfils its requirements. The client should have a clear conception of its objectives, but those of the design-and-build firm may at times conflict with those of the client.

If, for instance, a problem needs resolving, in which the achievement of the best design solution conflicts with the method of construction, the design-and-build firm may seek to solve the problem by opting for ease of construction at the expense of the best design solution. Other similar problems may occur, for instance economy versus form of construction, speed versus construction method, etc. The client will be in a position to resolve situations to its benefit if it

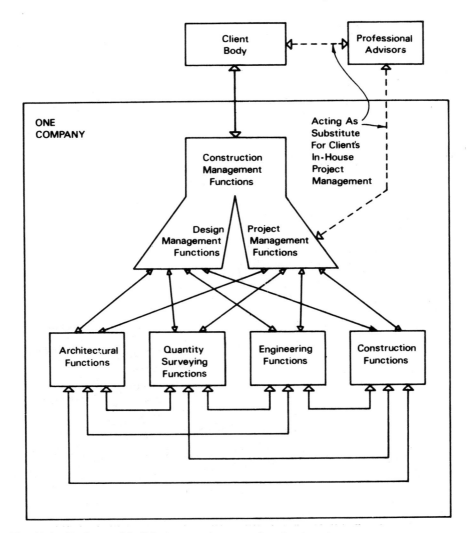

Fig. 11.4 Design-and-build structure (construction dominant).

has sufficient in-house expertise to understand the issues and the appropriate contract conditions that allow the client to act to produce a result to its benefit. If the client has not, then professional advice upon which to act will be needed.

Professional advisors in this capacity would act as a substitute for the client's in-house project management team as shown on Figs 11.4 and 11.5. Integration between the professional advisors and the management structure of the design-and-build firm would have to be carefully designed to ensure that it had effective lines of communication and authority, which would have to be made explicit in the contract conditions, e.g. whether they are in an advisory or executive position.

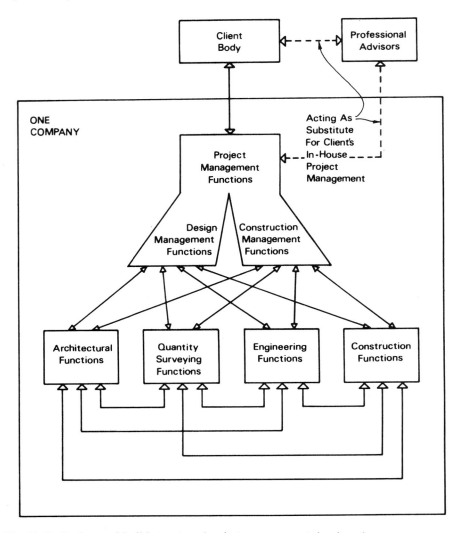

Fig. 11.5 Design-and-build structure (project management dominant).

Naturally, this implies that the professional advisers should be closely involved in drafting the conditions of contract and in establishing the project.

An extension of the design-and-build approach that has gained more acceptance in recent years in the UK and is more widely used abroad is that, rather than arranging a contract with one selected contractor, competition takes place for both design and price. Alternatively, an upper limit on the price may be fixed so that essentially competition is based primarily on design. The management arrangements with the successful bidder would be identical to those discussed earlier for one selected firm but either the client's in-house team or professional advisors would need to evaluate the submissions against the client's criteria to advise on the bid to be accepted. In such an arrangement, bidders would be provided with details of the client's requirements, which would normally include a performance specification. These details would form the basis of the criteria against which bids would be judged. Therefore integration of the professional advisors with the client or integration within the client's organisation with its own in-house team is of paramount importance in drafting the client's requirements for the bidders and in evaluating the bids.

Organisation matrix

At their simplest level, organisation structures of projects can be seen to consist of three major components – the client, the design team and the contractor. The client's experience of construction, the organisation of the design team and the method of appointment of the contractor will have a fundamental influence on the effectiveness of the project organisation. Examples of the range are:

(a) Client
  (i) No construction expertise, a senior manager liaises between client and project team.
  (ii) In-house expertise available, project manager appointed within client organisation.

(b) Design team
  (i) Conventional organisation
  (ii) Non-executive project manager.
  (iii) Executive project manager.

(c) Contractor's appointment
  (i) Selective competitive tender.
  (ii) Two-stage competitive tender.
  (iii) Competitive serial tender.
  (iv) Negotiated tender.
  (v) Management contract.
  (vi) Separate trade contracts.
  (viii) Design-and-build (overlaps with (b) above).

These classes are not exhaustive, but they represent the more common classifications and themselves produce a $2 \times 3 \times 7$ matrix, giving 42 alternative arrangements.

Each arrangement will present certain advantages and disadvantages and should be selected for use in circumstances that suit the particular project. The features of each arrangement are as summarised briefly in Table 11.1 and whilst they are drafted in terms of building projects, the structures are also relevant to civil engineering.

Table 11.1   Matrix of project organisation structure.

Client	Design team	Contractor's appointment
(1) No construction expertise	Conventional	Selective competition

Comments
This is the traditional arrangement and relies upon the architect as designer and manager, with the client having a limited management contribution to its project. The contractor is not integrated into the design phase. Suitable for relatively simple projects in terms of both complexity of construction and environment for which client's requirements are clear.

(2) No construction expertise	Conventional	Two-stage, competitive

Comments
Opportunity to use contractor's expertise during part of design phase depending upon when first-stage tenders are initiated. Gives opportunity to speed up programme by overlapping construction and some design work. Appears to be a half-hearted attempt to integrate the contractor.

(3) No construction expertise	Conventional	Competitive serial

Comments
As (2) above but used where a number of similar buildings are required for the same client.

(4) No construction expertise	Conventional	Negotiated

Comments
This arrangement enables the contractor to be integrated at a very early stage in the project. Relies upon the architect as designer/manager being prepared to use and trust the contractor's expertise. Suitable for complex projects and/or environments.

(5) No construction expertise	Conventional	Management contractor

Comments
In management terms, similar to (4) above but allows for competition for construction work. Still relies upon the architect as designer/manager being prepared to use and trust the contractor's expertise. The management contractor is acting as another consultant with responsibility for arranging and organising subcontractors and may be more acceptable to other members of the design team in this capacity. Potential conflicts between architect acting in a management capacity as well as being designer and the management role of the management contractor. Suitable for complex projects and/or environments. Illustrated in Fig. 11.6.

(6) No construction expertise	Conventional	Separate trade contracts

Comments
Potentially high in integration by direct contact between the design team, particularly the architect and subcontractors on site which could be extended into the design phase. Reduces differentiation of major management components by eliminating a separate general contractor function. Requires an architect leader who is strongly management orientated and objective about the often conflicting demands of design and construction. Illustrated in Fig. 11.7.

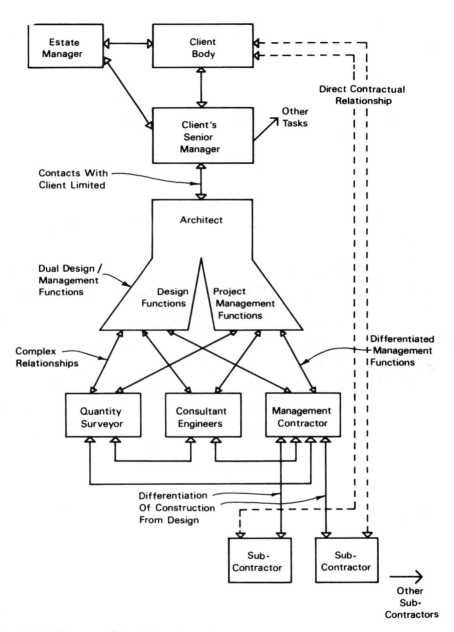

Fig. 11.6   Diagram of type 5 structure.

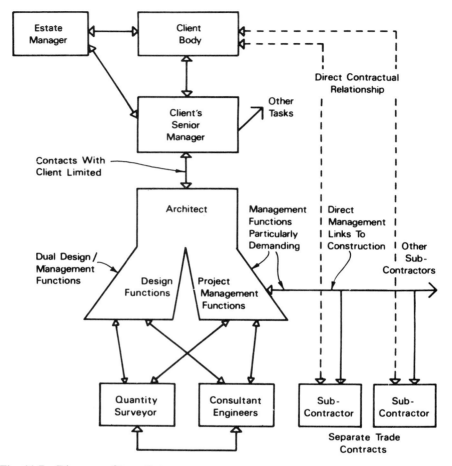

Fig. 11.7   Diagram of type 6 structure.

Table 11.1   (continued)

Client	Design team	Contractor's appointment
(7) No construction expertise	Conventional	Design-and-build

Comments

This arrangement would not be used as described. The conventional design team would act as consultants to the client to monitor the design-and-build contract. As the client has no in-house expertise, some professional advice would be required to assist the client's representatives in dealing with the design-and-build contractor. The conventional design team in this context would probably consist of one or two professional advisors (or a firm). Integration should be strong within the design-and-build firm but the potential weakness is in the integration with the client when it has no in-house expertise, even though it has appointed advisors. Illustrated in Figs 11.4 and 11.5.

Table 11.1   (continued)

Client	Design team	Contractor's appointment
(8) In-house expertise	Conventional	Selective competition
(9) In-house expertise	Conventional	Two-stage competition
(10) In-house expertise	Conventional	Competitive serial
(11) In-house expertise	Conventional	Negotiated
(12) In-house expertise	Conventional	Management contractor
(13) In-house expertise	Conventional	Separate trade contracts
(14) In-house expertise	Conventional	Design-and-build

Comments

This group is similar to (1) to (7) above except that the client has in-house building expertise available and appoints a project manager from within its own organisation.

The result in each case is that integration between the project team and the client should be closer, provided that the client's own internal integration is effective. A member of the client's organisation appointed project manager should have more time to devote to the project than if a senior manager were doing this job in addition to normal work. The demands on the architect to exercise this management role effectively are likely to be greater as a result of the pressure exerted by the project manager.

Approaches to appointing the contractor and integrating the contractor into the team other than by selective competitive tender are more likely to be adopted because of the influence of the client's in-house project manager.

Alternative (14) is unlikely to be used as the in-house project manager should normally have the expertise and capability to monitor the design-and-build contract. Alternative (13) is illustrated as an example in Fig. 11.8.

Client	Design team	Contractor's appointment
(15) No construction expertise	Non-executive project manager	Selective competition
(16) No construction expertise	Non-executive project manager	Two-stage competition
(17) No construction expertise	Non-executive project manager	Competitive serial
(18) No construction expertise	Non-executive project manager	Negotiated
(19) No construction expertise	Non-executive project manager	Management contractor
(20) No construction expertise	Non-executive project manager	Separate trade contracts
(21) No construction expertise	Non-executive project manager	Design-and-build
(22) In-house expertise	Non-executive project manager	Selective competition
(23) In-house expertise	Non-executive project manager	Two-stage competition

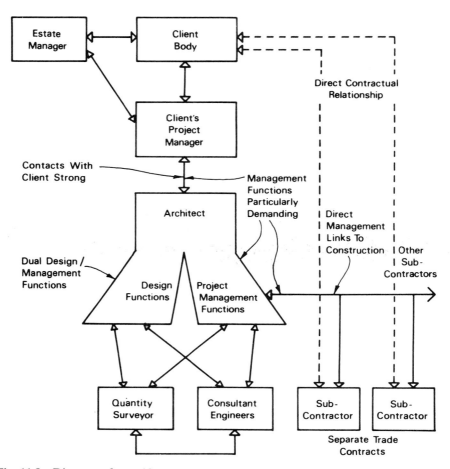

Fig. 11.8   Diagram of type 13 structure.

Table 11.1   (continued)

Client	Design team	Contractor's appointment
(24) In-house expertise	Non-executive project manager	Competitive serial
(25) In-house expertise	Non-executive project manager	Negotiated
(26) In-house expertise	Non-executive project manager	Management contractor
(27) In-house expertise	Non-executive project manager	Separate trade contracts
(28) In-house expertise	Non-executive project manager	Design-and-build

Table 11.1 (continued)

Client	Design team	Contractor's appointment

Comments

This group corresponds with (1) to (14) above, except that a non-executive project manager is appointed within the project team. As discussed earlier, someone in this position will fill a co-ordination and communication role without authority for executive functions. Provided that the role is recognised and accepted by the other team members, the administration of the project should benefit. But as the decision-making structure and authority pattern remain unaltered, the effect on the management of the project is unlikely to be significant and the comments given for (1) to (14) will apply. The non-executive project manager is likely to be dominated by the designer and in the cases where the client appoints an in-house project manager, he or she will be particularly easily overridden. Alternative (25) is illustrated in Fig. 11.9.

(29) No construction expertise	Executive project manager	Selective competition

Comments

Management and design responsibilities are split, which should allow the project manager to concentrate upon management of the project both within and between the design and construction teams and with the client. Probably the most effective way of improving the management of what is still really a conventional structure, suitable for complex projects and/or environments, where it is necessary for the contract to be awarded competitively.

(30) No construction expertise	Executive project manager	Two-stage competition

Comments

Comments as for (2). Opportunity is presented for the project manager to ensure that the contractor is properly integrated and makes a contribution to the design phase. It will be up to the project manager to time the first phase so that the contractor's contribution is maximised. Illustrated in Fig. 11.10.

(31) No construction expertise	Executive project manager	Competitive serial

Comments

As (30) above, but where a number of buildings are required for the same client.

(32) No construction expertise	Executive project manager	Negotiated

Comments

This arrangements allows the project manager to establish a tightly integrated team from the very early stages of the project. Allows the project manager to appraise the contribution from all members of the project team objectively. Suitable for very complex projects and/or environments.

(33) No construction expertise	Executive project manager	Management contractor

Comments

In management terms similar to (32) above, but allows for competition for construction work. As the management contractor is acting as another consultant, the project manager may be able to integrate the management contractor with less constraint than may be the case with (32). Suitable for very complex projects and/or environments. Illustrated in Fig. 11.11.

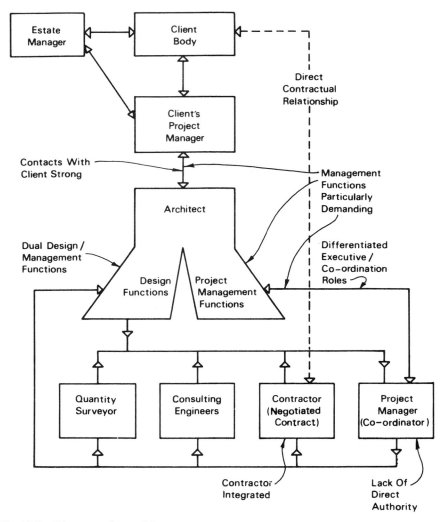

Fig. 11.9   Diagram of type 25 structure.

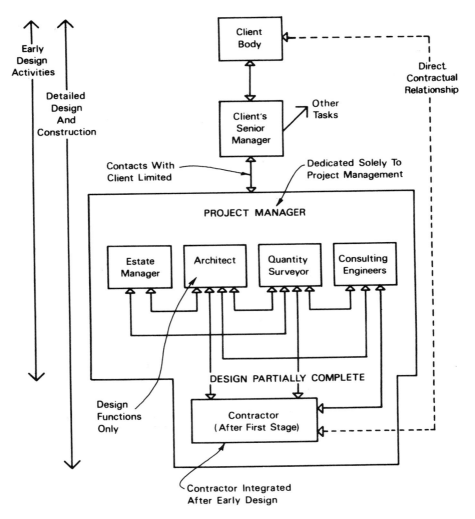

Fig. 11.10 Diagram of type 30 structure.

Table 11.1 (continued)

Client	Design team	Contractor's appointment
(34) No construction expertise	Executive project manager	Separate trade contracts

Comments
Potentially high in integration by direct contact between the design team, particularly the project manager and the subcontractors. Reduces differentiation of major management components by eliminating a separate general or management contractor function. The use of a project manager who is strongly management orientated and objective about the often conflicting demands of design and construction should provide the foundations for a successful project. Suitable for complex projects and/or environments.

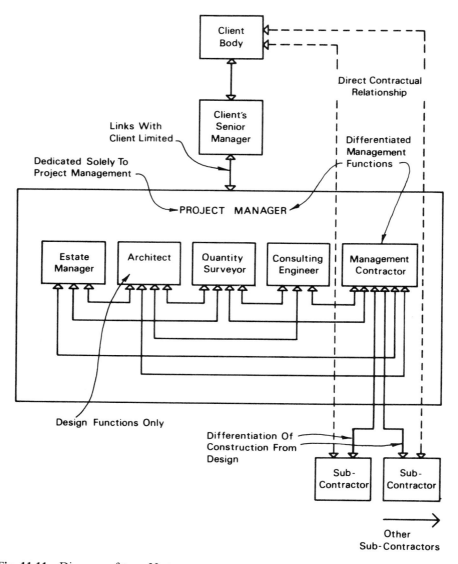

Fig. 11.11   Diagram of type 33 structure.

Table 11.1 (continued)

Client	Design team	Contractor's appointment
(35) No construction expertise	Executive project manager	Design-and-build

Comments
In this arrangement the executive project manager would not have a supporting design team but would act as the client's representative in monitoring and controlling the design-and-build contract. Integration should be strong within the design-and-build firm. Integration between the client and project manager and between the project manager and the design-and-build firm would depend to a large extent on the relationships established in the formal contract. The project manager should be involved in establishing the contract in which his or her authority should be clearly established.

Client	Design team	Contractor's appointment
(36) In-house expertise	Executive project manager	Selective competition
(37) In-house expertise	Executive project manager	Two-stage competition
(38) In-house expertise	Executive project manager	Competitive serial
(39) In-house expertise	Executive project manager	Negotiated
(40) In-house expertise	Executive project manager	Management contractor
(41) In-house expertise	Executive project manager	Separate trade contracts
(42) In-house expertise	Executive project manager	Design-and-build

Comments
This group is similar to (28) to (35) above, except that the client has in-house building expertise available and appoints a project manager from within its own organisation in addition to the executive project manager of the project team. The result in each case is that integration between the project team and the client should be closer, provided that the client's own internal integration is effective. A member of the client's organisation appointed project manager should have more time to devote to the project than a senior manager who may also have to do routine work. The use of the two project managers should strengthen the integration with the client to the extent that this group of arrangements should represent potentially the strongest management-dominated organisational structure. Alternative (41) is illustrated in Fig. 11.12. In particular the use of structures that allow the contractor to be integrated into the team (39, 40, 41) has the potential for full integration and allegiance to objectives of even the most complex projects, the facility for implementing the most rigorous control mechanisms and the opportunity to take the most appropriate advice before decisions are taken.

Whilst it has been recognised that the selection of procurement systems by clients has become increasingly complex (Masterman & Gameson 1994), one of the key issues in the choice of organisation structure lies in the trade-off between

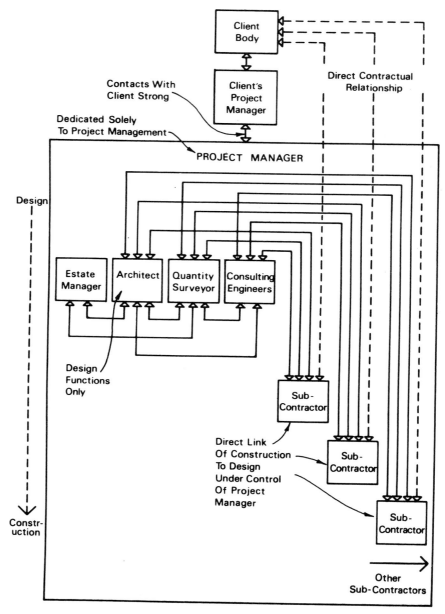

Fig. 11.12   Diagram of type 41 structure.

the apparent competitiveness of the bid price for the construction work and the early involvement of the contractor in the project team. Although competition for a construction contract may appear to provide the client with the lowest price, it is achieved only at the cost of a potentially less integrated project team, as the contractor cannot be brought into the design team sufficiently early to influence the 'buildability' of the design. The effect of this could be a longer construction period and a higher price because of the difficulty of construction of the proposed design. Hence the competitiveness of the bid may be more imagined than real. In the interests of the total economy of the project, a fully integrated team throughout the whole process may be more beneficial to the client's interests. This factor will be most important in selecting the appropriate structure of the major components of the project organisation to suit the particular circumstances of the project and the needs of the client.

Chapter 12
Analysis and Design of Project Management Structures

Need for analysis and design

The mainly theoretical scenario developed so far provides the concepts necessary for analysing project organisations and the basis for a structured approach to their design. Such theory is all very well but it needs translating into techniques that are useful in practice and make a positive contribution to improving the effectiveness of the management of projects in the real world. For analysis, such techniques should be capable of mapping what actually happened on projects. When used for design they should be capable of showing clearly what is expected to happen if the project is to be managed effectively. The predominant need is to design structures that allow people to work together effectively, but there is also the parallel need to develop structures that enable the use of appropriate project management techniques.

The increasingly explicit recognition of project management in its own right rather than as something subservient to other professional skills has helped to generate a range of techniques and tools for project control, e.g. critical path networks, line of balance, etc. The opportunity and the will to employ such techniques depend upon a receptive management structure and the effective organisation of contributors. A management structure led by people whose priority is management should result in more widespread use of project control techniques and ensure that contributors to the project are organised to maximise the benefits of such techniques. But what is often overlooked is that it is the responsibility of those exercising project management skills on behalf of clients to design organisation structures appropriate to particular projects and their environments so that the right skills and techniques are used at the right time. This is particularly difficult to achieve in construction as the structures normally used are predominantly conventional and reflect the juxtaposition of traditional professional roles. This tends to inhibit innovation, with the result that the industry and professions are slow to apply new ideas and techniques.

Criteria

Techniques for analysing and designing organisational structures should make clear the following aspects of the way in which the project is organised:

- the operating system
- the managing system
- the relationship of people in the organisation and their interdependency
- the roles of the people in the organisation
- the position of the decision points and their status, e.g. key, operational
- the contribution of people to each decision and their relationships in arriving at decisions.

An approach that exposes these aspects will give great visibility to the way in which projects are organised. It will show clearly who did what and, perhaps more significantly, who did not. Compared with the traditional pyramidal organisation chart, such an approach would, if used for organisation design, be a dynamic representation of what should happen on a project, or, if used for analysis, what really did happen, rather than being a simplistic, static statement of who is whose boss, without any attempt to relate the people to the project activities and to each other.

Techniques

Two techniques go some way to satisfying the criteria: *transformed relationships evolved from network data* (TREND) and *linear responsibility charting* (LRC). In principle, both are based upon a network approach but expose the relationships of the people involved in the project rather than the time relationship of activities, which is the normal use of network techniques.

TREND (von Seifers 1972; Bennigson & Balthasas 1974) was developed to analyse organisations in a study of temporary management structures. The categories of information required were identified as:

- What tasks are to be performed?
- Who is responsible for each task?
- How do the tasks interrelate?
- What is the interaction pattern of the participating departments?
- Which tasks and departments are more critical to successful completion?
- What is the nature and location of the uncertainties involved?

It was recognised that much of this information could be obtained from a network, but that to answer the last two items of the above list a PERT network with three time estimates for activities would be required. The technique was used for examining what the study identified as the two initial variables affecting organisation design: the nature of the tasks and the resource group interdependencies, which were taken as subsuming the above list of questions.

Using TREND it is possible to identify the resource groups responsible for activities contained in the network and the types of interdependencies between resource groups generated by the project. The nature of the tasks was assessed from the level of uncertainty associated with activities, and the critical nature of activities, both of which were represented by the three time estimates of PERT networks.

TREND has the potential to provide a powerful and sophisticated tool for project organisation design where a high level of uncertainty needs to be managed. However, ideally it requires a PERT network with three time estimates or information to this level of detail. In the original study carried out in the USA it was found that there were very few projects available with this degree of detail in networks. In fact there were only a limited number of projects for which there was a network of any kind. As a corollary the technique was found to have limited application to long-duration projects with much aggregation of the activities in the project plan. As most construction project plans are of this type, the application and usefulness of TREND may be limited until such time as the construction industry adopts more sophisticated planning techniques.

An alternative approach, which does not require extensive networks to be available and which incorporates within it a technique for building up organisation networks at various levels of detail, has been developed from linear responsibility charting. This first appeared in the mid 1950s (Anon 1955) but does not appear to have been further developed until about 1975, by Cleland and King (1983). Linear responsibility charting originated as an improvement upon the pyramidal organisation chart so that it shows who participates and to what degree when an activity is performed or a decision made.

Linear responsibility analysis

It was subsequently discovered that LRC could serve as a tool for organisation design and analysis since it can be made to display system interfaces and interrelationships. It was then further advanced into the linear responsibility analysis (LRA) technique in 1980. This method allowed the degree and quality of differentiation, integration and interdependency within an organisation system to be identified and was applied to building projects (Walker 1980). Decision points were overlain on the LRA to show the combination of contributors to decisions and their roles.

Although arising from general management thinking, LRA is particularly useful for project management. Its effectiveness lies in its ability to expose both the decision points in the process and the way in which the people in the operating and managing systems are arranged in relation to those decision points. It also allows identification of the activities of the people in the managing and operating systems and of the relationships between them. Thus LRA gives visibility to the arrangement and integration of contributors bringing forward propositions upon which decisions are based. An understanding of how this process works is fundamental to the success of a project organisation and LRA makes a significant contribution.

Starting with the linear responsibility chart (LRC) from which the linear responsibility analysis (LRA) originated, the best way to understand the approach and how information is presented is to examine a typical chart such as that for the design of a building project, shown in Fig. 12.1.

The LRC illustrated uses eleven symbols (see page 232 for detailed description) to indicate different types of relationship that may exist between any job position in the organisation structure and any task to be performed. The job positions are listed along the horizontal axis of the matrix and the tasks to be carried out are listed down the vertical axis. In the square where a job position and a task meet, the relationship is indicated by inserting the appropriate symbol. If a job position has no relationship with a task, the square is, of course, left blank.

For example, for Task 6, 'Contractual Proposals', the partner of the quantity surveying practice prepared the proposals and in doing so consulted the partner of the architectural practice, the managers of the services engineering and structural engineering departments and the senior partner of the engineering practice. This was done so under the management of the project manager who was concerned with boundary control, and with maintenance and monitoring. The project manager finally recommended the proposals to the client's project engineer. The latter approved them as well as being consulted for instructions and advice during the work and exercising general oversight.

A great benefit of the LRC is the virtue of presenting much in little space in a dynamic form, but it is much more than this. It gives an overall perspective of a project organisation structure which brings to life relationships in a way which more conventional approaches such as pyramidal organisation charts do not. It also gives a basis for further development of more sophisticated and effective techniques of organisation design.

The degree of detail of the tasks selected for the vertical axis is under the control of the designers of the organisation, either as a result of the information which they current have available or as a result of the particular aspect of organisation design with which they are concerned. Similarly, the job positions can be in the range from individual people to whole departments or firms. The symbols likewise can be selected for the particular purpose. The inherent flexibility of the chart is very valuable to designers in that it allows them to orientate their work to the particular problem they wish to study.

Cleland and King (1983) enhanced the LRC by visualising it as an input – output device, as shown in Fig. 12.2. The input is the person in the job position with the 'does the work' relationship to the task, and the output is the completion of the task. The inputs – outputs (or 'does the work' relationships) are then transformed into schematic form, as shown in Fig. 12.3. The connected boxes containing the job position which 'do the work' and the tasks they carry out form

Legend

- ○ Did The Work
- ▲ Approves
- ▶ Recommends
- ● General Oversight
- ◆ Direct Oversight
- △ Boundary Control
- □ Monitoring
- ◇ Maintenance
- ■ Consultation - Gave Instructions And Information
- ▽ Consultation - Gave Advice And Information
- ⊛ Output Notification Mandatory

MAJOR TASK	Parent Company	Local Board Of Directors	Responsible Director	Project Engineer	Department Managers	Specialist Staff	Workforce	Senior Partner	Project Manager	Manager Structural Engineering	Job Structural Engineer	Manager Services Engineering	Job Services Engineer	Partner Architect	Job Architect	Partner Quantity Surveyor	Job Quantity Surveyor
	Client →									*Consultants — Engineers* →				*Architect* →		*Q.S.* →	
1 Identify Need For Project		●▲	◆▶	△□◇	○												
2 Define Outline Requirements		●▲	◆▶	△□◇	○	▽		▽									
3 Establish Budget Estimate		●▲	◆	△○	○			▽									
4 Presentation For Inclusion In 5 Year Plan	▲	●▼	◆	△○													
5 Programme Proposals				▲■●				◆▽	▽△◇	▽		▽		▽		▽	
6 Contractual Proposals				▲■●				◆▽	▽△◇	▽	○	▽	○	▽	○	○	
7 Spatial Proposals				▲■●			▷	◆▽	▽△◇	◆	○	◆	○	◆	▷	▷	▷
8 Technical Proposals (Structural)				▲■●		▷	▷	◆▽	▽△◇	◆	◆	▽	▷	▽	▷	▷	▷
9 Technical Proposals (Services)				▲■●		▷	▷	◆▽	▽△◇	▽	▷	◆	▷	▽	▷	▷	▷
10 Technical Proposals (Architectural)				▲■●		▷	▷	◆▽	▽△◇	▽	▷	▽	▷	◆	○	▷	▷
11 Financial Proposals				▲■●				◆▽	▽△◇	▽	▷	▽	▷	▽	▷	◆	○
12 Consolidate Brief		◀	▶	▼●				◆▽	▽△◇	▽	▷	▽	▷	▽	▷	▷	▷
13 Capital Expenditure Presentation	◀	▶	▶	▼○													
14 Programme Details				▲▽				●	▽△◇	▽		▽		▽		▷	▷
15 Working Drawings				▲▽■				●	▽△◇	◆	○	◆	○	◆	○	▷	▷
16 Technical Details (Structural)				▲▽■				●	▽△◇	◆	◆	▽	○	▽	▷	▷	▷
17 Technical Details (Services)				▲▽■				●	▽△◇	▽	▷	◆	○	▽		▷	▷
18 Technical Details (Architectural)				▲▽■				●	▽△◇	▽	▷	▽	▷	▽	▷	▷	▷
19 Contract Details				▲▽				●	▽△◇	▽	▷	▽	▷	▽	○	◆	○
20 Contract Documentation				■				●	▽△◇					◆		◆	○

Fig. 12.1 Typical linear responsibility chart.

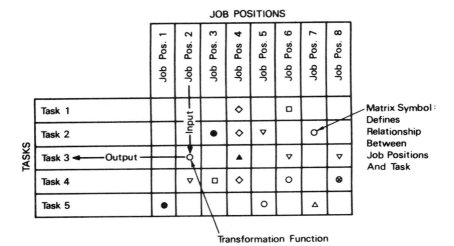

Fig. 12.2 LRC matrix (showing input–output applications).

the operating system through which the project is achieved. To this schematic are added the job positions of those who are involved in managing the process together with the symbols representing their specific management functions in connection with each task. These are placed in loops above the task boxes and are known as the control loops. The control loops represent the managing system of the project. People in relationships other than in the operating or managing systems, mainly in consulting roles, are then added as shown. When completed for the whole project, the schematic LRC shows the way in which the tasks are connected and how people act and interact within the organisation in carrying out the project.

This presentation clarifies the operating system (the linked tasks) through which the project is carried out, and the managing system (the job positions in the control loops) which controls the operating system and the relationships of others who contribute.

A further development of the LRC with the title linear responsibility analysis involves the addition of further systems information to give a sharper view of the system as a whole. This total systems picture of the project enables an objective assessment to be made of the level of integration necessary for the system to work effectively. It also enables weaknesses in the system to be identified. If necessary,

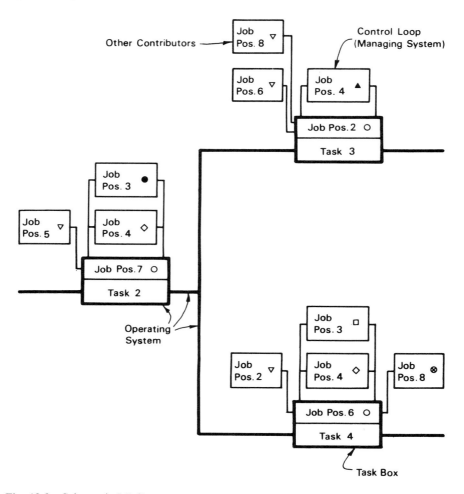

Fig. 12.3   Schematic LRC.

the system can be redesigned at those points or alternatively particular attention
can be paid to integration. This development, as illustrated in Fig. 12.4, adds the
following to the schematic LRC:

(a) *Interdependency*: The type of interdependency between tasks is shown,
    sequential interdependency by a solid line and reciprocal by a broken line.
    The type of interdependency influences the degree and type of integration
    required. Reciprocal interdependency requires greater integrative effort than
    sequential and needs a more flexible approach. Projects with a high pro-
    portion of reciprocal interdependencies will require great skill and effort in
    integration.

(b) *Differentiation*: The causes of differentiation between the contributors are shown by symbols representing the various types. As discussed in Chapter 7, differentiation is caused by:
  (i) *technology* (or skill – the technical demands of the job someone does, which determine the way in which work is divided between groups of people (T_1);
  (ii) *territory* (or location) – the geographical distance between people (T_2);
  (iii) *time* – the sequence of people's work on the project (T_3).

The factors creating differentiation can also be reinforced on construction projects by *sentience*. A sentient group is one to which individuals are prepared to commit themselves and on which they depend for emotional support. On the LRA it is identified as:
  (i) sentience arising from profession allegiance (S_1);
  (ii) sentience arising from both professional allegiance and allegiance to a firm (S_2).

The greatest degree of differentiation that can be shown on the LRA between contributors is therefore represented by the symbols T_1. T_2, T_3, S_2. The minimum is when there is no differentiation present in the relationship.

An examination of the degree of differentiation shown on an LRA will indicate the amount of integrative effort that will be necessary and in particular when and where on the project integrative effort is likely to be especially important. Conversely, areas of the project that should require relatively low levels of integrative effort will be revealed.

(c) *Decision points*: The decision points at the various levels of the hierarchy (primary, key and operational) are overlain on the LRA. When designing an organisation structure, the decision points have to be anticipated. This is part of the planning process and requires the close collaboration of the client. A particular skill is to ensure that the arrangement of contributors is appropriately designed prior to each particular decision. When used for analysis, the arrangement of the contributors relative to the decisions made is exposed for examination. As referred to previously, the decision points provide feedback opportunities which should be taken for the project to be completed successfully. They can be indicated on the LRA together with details of the control against which the state of the project's development has to be measured.

(d) The systems and sub-systems that make up the total project system can now be identified. They will overcome conventional professional boundaries and enable the project participants to visualise the project in terms of interrelated tasks and people rather than in terms of professional compartments.

An example of a small part of an LRA of a completed project used for postmortem purposes is shown in Fig. 12.5.

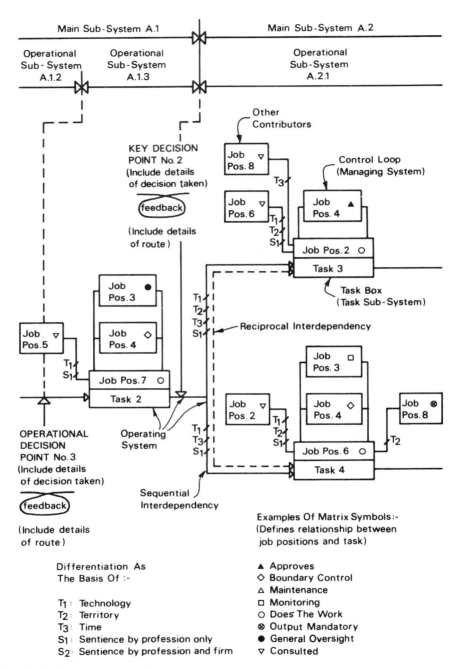

Fig. 12.4 Linear responsibility analysis in principle.

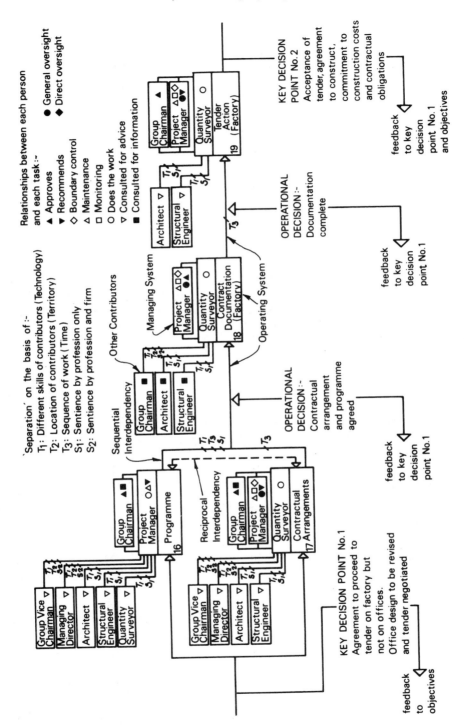

Fig. 12.5 Application of linear responsibility analysis.

Matrix symbols

The symbols define the way in which job positions relate to tasks and are selected and defined to suit the particular needs of the project. Each relationship can be classified into one of four categories:

  (i) a transfer function of input into output within the operating system
 (ii) a control loop function concerned with managing the operating system
(iii) a contribution of input to a task, external to the operating system
(iv) a receipt of output from a task, external to the operating system.

Examples of relationships appropriate to the construction process, derived from Cleland and King (1983) are now described.

(a) Transfer function
'Does the work': This is where inputs to tasks are transformed into outputs from tasks in accordance with instructions. It is the juncture of managing and operating systems where the output is transferred under the control of the managing system to be input to the next task. This relationship appears in each task box, and the total of the task boxes defines the operating system and those involved in it. It is the relationship in which professional skills are directly applied to the project, e.g. designing, constructing, preparing documentation, etc.

(b) Control loop functions

(i) Approves: This constitutes the final control loop function. The person in this executive relationship has the authority to approve the output of tasks for use in further tasks on the project. Normally, it is to be expected that the client will retain approval power for tasks directly affecting primary and key decisions, and that the member of the project team responsible for project management will approve those affecting operational decisions and other output.

(ii) Recommends: The person in this relationship is charged with the responsibility for making recommendations for approval of the output of tasks. The member of the project team responsible for project management will usually make recommendations to the client for approval. If the project manager is required to approve the output of the task, then some other member of the project team would normally make the recommendation.

(iii) General oversight: This is the broadest administrative control element and the source of policy guidance. The person in the direct oversight relationship responds to the wishes of the person in this relationship. The person in the general oversight relationship will not himself be exercising the skills of a task over which he has oversight. The primary role is to furnish policies and guidance of a scope that permits as much decision-making flexibility as possible within a task in arriving at the output. An example would be the role of a senior partner of a

professional consultancy engaged to manage the project. That person would not normally be working within the operating system but would be managing the firm's input and on these occasions would appear in the control loop as part of the managing system. The omission of the relationship from a control loop indicates that the task was assumed not to involve questions of policy.

(iv) Direct oversight: This is the administrative control element immediately below the 'general oversight' relationship. Although having no specific project functions, the person in this relationship has, and will exercise when necessary, the skills demanded by a task over which he or she has oversight. The person is seen by others involved in the project to be maintaining a presence close to project activities. The omission of this relationship from a control loop indicates that the task was of such a routine nature that direct supervision was not necessary. An example could be a partner or senior assistant in a professional consultancy who is leading a team of people who are actually doing the work required by a particular task. The frequency of the appearance of this symbol on an LRA depends upon the level of detail of the LRA. Whereas it may not appear on an LRA prepared at a strategic level as it would be subsumed within the 'does the work' relationship in the task box, it would appear frequently if an LRA were prepared at a greater level of detail, say to plan the activities of one critical aspect of the project organisation which required close examination of a particular professional consultancy's contribution.

(v) Boundary control: When this appears in a control loop it indicates the specific control activity of ensuring functional compatibility within the task for which it appears and between it and other tasks. The person in this job position does not normally also have an administrative role. In addition, this relationship is concerned with relating the total system to its environment. It is a project management function as it is concerned solely with integrating the tasks and hence the people working on projects, and with ensuring that the tasks and the total system respond to changes in the project's environment. Omission of this relationship from a control loop indicates that the task is undertaken independently of other tasks in the operating system, which is an unlikely occurrence on a construction project.

(vi) Monitoring: This is the specific control activity of intra-task regulation concerned with checking prior to output to ensure that a 'does the work' activity is achieving its purpose. Omission of this relationship from a control loop indicates that it was not necessary to carry out such checks because there was total confidence that the activity would achieve its purpose.

Monitoring is a project management function. Although this activity may be considered to be the responsibility of the firm providing a particular contribution to the project, it is nevertheless the responsibility of project management to ensure that all aspects of the system are performing satisfactorily. Even if a particular task is receiving 'direct oversight' from a senior member of the firm providing the contribution, it is to be expected that the person providing project management would exercise a monitoring function over all activities.

(vii) Maintenance: This is the specific activity of ensuring that a 'does the work' activity is being maintained in an effective state, both quantitatively and qualitatively, so that it is capable of achieving its purpose. Omission of this relationship from a control loop indicates that it was not necessary to maintain the 'does the work' activity. Again, this is a project management function concerned with making sure that the right number of people of the appropriate quality are employed on a task. As with the last relationship, even though the managers directly associated with the task may also consider that this is their function, it is also the responsibility of whoever is managing the total project on behalf of the client.

(c) Contributions to input

(i) Consultation – gave instructions and information: This is an input of instructions and information to a 'does the work' activity and does not therefore appear in the control loop. Boundary control activities should ensure that the necessary people are placed in this relationship to appropriate tasks.

(ii) Consultation – gave advice and information: This is comparable to the last relationship but advice (rather than instructions) and information are input to a 'does the work' activity. Again, boundary control activities should ensure that the necessary people are placed in this relationship to appropriate tasks.

(d) Receipt of output
Output notification mandatory: This is placed in the output of a task when it is essential that the person in this relationship with a task receives timely information concerning a task output. The concept of this relationship is one of passive transmission of information.

Benefits

An LRA gives great visibility of how people work or do not work together on a project. In particular, it makes clear:

(i) *The sequence of the tasks and the effectiveness of the process in relation to the decision points and the actual decisions taken.* The tasks undertaken before a decision is made and the skills and status of the people contributing to the tasks and the decision will be clearly exposed.
(ii) *The type of interdependency, which identifies the way in which the tasks are related.* Inappropriately drawn interdependencies, e.g. sequential, which should have been reciprocal, are normally ineffective and most difficult to integrate (for instance, if the structural engineer and the quantity surveyor

simply respond to the architect rather than each influencing the other's decisions).

(iii) *Who contributes to each task and with what status and what relationship to others.* More importantly, perhaps, it shows who does not contribute or who contributes in the wrong relationship. The LRA shows the degree of differentiation between contributors and hence the degree of integration required. The amount of differentiation is reflected quantitatively by the number of links between tasks and between job positions within tasks. The simpler the project the fewer links there will be. The links between tasks represent the differentiation that has to be integrated in managing the output of the tasks to realise the project. The links between job positions within tasks give the differentiation to be integrated in achieving the output of each task.

A qualitative measure of differentiation is given by identifying the proportion of each permutation of the factors of differentiation (technology, territory, time and sentience). Theoretically, it is possible to have any permutation of differentiation factors (T_1, T_2, T_3, S_1, S_2), but in practice the configuration of the contributors limits the range which occurs on any project. For example, within a task the various professions in a multidisciplinary practice in which all members are located in one office and can only have a differentiation of T_1, S_1 (technology and sentience by profession).

It is to be expected that maximum integration will only be possible if an appropriate pattern of consultation has been established between the contributors. Such a pattern of consultation is demonstrated by all relevant job positions being in the 'gave advice' or 'gave information' relationships with each task.

Similarly, the degree of differentiation of the system will be a function of the pattern of consultation established. It is to be expected that differentiation will be greater as the number of contributors to each task increases as a function of the complexity of the project. This will, therefore, demand greater intensity of integration.

(iv) *Who exercises what management role as shown by membership of the control loop and who does not, but should do.* The control loops show the degree of continuity between tasks, particularly at decision points, and indicate the level of integration between tasks. The degree of integration exercised over the contributors to each task is identified by the activities of the members of the control loop.

Ideally the control loop composition should show continuity of membership. Normally, the member of the project team responsible for project management and a representative of the client organisation should appear in all control loops and should consistently exercise the same functions. Interruptions of continuity identify potential weaknesses in the integrative functions of the managing system. Interruption of the managing system between tasks in this way occurs particularly when the person in the boundary control relationship does not appear in the same relationship in successive control loops. Such discontinuity may be especially significant at decision points.

Duplication of management functions within a control loop may occur and represents split responsibility between members of the managing system. As well as being undesirable in itself, it also adds to the complexity of the managing system and can impair its effectiveness.

Control loop composition identifies the level of separation of the managing and operating systems. If this separation is not complete, a potential weakness in integration may occur as the objectivity of the person occupying dual roles may be in doubt. If the managing and operating systems are totally undifferentiated, there would be no control loops and all task boxes on the LRA would be occupied by the same job position. If the systems are totally differentiated, none of the task boxes would be occupied by any job position appearing in any control loop.

(v) *The integration of the client's representative with the contributors to the project.* The degree to which the client's representative is involved in the project is demonstrated by his or her presence in the control loops or other relationships. In addition it shows the type of relationship maintained with the tasks, e.g. approval power, general supervision, etc. Consistency of the client's representative is also shown.

Integration of the client's representative and the project team takes place within the tasks and between them. Within a task it is shown when the client's representative appears in the control loop for the task together with the member of the project team responsible for project management. Integration of the client's representative occurs between tasks when the same client's representative appears in the control loops of successive tasks together with the same member of the project team.

Interpretation and use

As can be seen, a particularly interesting aspect of any analysis or design is the composition of the control loops. On projects that use a project manager from outside the client organisation but with close involvement of the client, it is to be expected that all control loops will consist of two members, the project manager and the client's representative. It will be interesting to observe the structure of the control loops if the client's representative has to involve other senior members of the client's organisation at various stages, particularly at key decision points. Continuity may break down and the functions exercised by the duplicated client's representatives in the control loops will generate complex relationships. Although such situations may be unavoidable, the use of LRA to design organisations will enable them to be identified in advance and allow steps to be taken to minimise disruption of the management of the project.

When the members of the project team do not organise themselves positively for managing the project, it is likely that their representative in the control loop will vary, giving lack of continuity between tasks and erratic relationships between the members of the control loops and the contributors who actually undertake the tasks. This is often the case where a senior partner of a professional consultancy is involved in securing a commission at the early stage of a project and then transfers responsibility for managing the project to another member of staff, particularly if that person then continues to maintain an ambiguous role in relation to the management of the project.

The level of detail chosen for the individual tasks and job positions will depend on the purpose for which the LRA is intended. This can range from using it to draw a broadly based map of the whole of a large project down to the design of the way in which a small section of the work is to be organised. Users can pursue the aspect of particular interest to them by selecting and defining the relationships they wish to design or study by the use of appropriately defined matrix symbols. An important advantage is that an LRA can be usefully employed at a level of abstraction suited to construction projects, which are invariably of long duration with a great deal of aggregation of the detailed activities in the project plan and hence in the data available. However, Morris (1994) considers that the technique can be very useful in a variety of project situations. Hughes (1989) has proposed a modification to the LRA which he terms '3R charts'. He believes that in order to handle the data for large complex projects, it is necessary to disaggregate the LRA chart into a series of inter-related components to separate the general from the detail. By doing so the difficulties of complexity in large projects is overcome.

Naturally, any LRA used for design should be capable of being updated and amended in the light of changing environmental conditions. As with any planning tool, it should not remain fixed if conditions change, but should be used to anticipate problems resulting from such changes so that they can be dealt with in the most effective manner.

It should, of course, be pointed out that LRA represents a structural approach to project organisation and that just because the appropriate relationship of contributors, operating system and managing system has been established does not mean that the people involved will work well together in the manner intended, and that the project will be successful. Even if the best organisation has been structured, fundamental aspects of success will be the quality of the skills brought to bear by the contributors, the pattern of authority and power, the leadership of the project team and the attitudes and personal relationships that develop on the project. All of these aspects will be significantly affected by the personal qualities of the managing member of the project team. Nevertheless, the LRA approach should result in an organisation being structured to suit the particular project as the technique demands that the operating system, together with the decision points, are identified before the organisation is fitted to it. That is, the design of the organisation should follow definition of the task to be accomplished. This should give the project manager a sound basis for harnessing the behavioural characteristics of the people involved in the project to the benefit of the client.

Project outcome

The major problem of evaluating the effectiveness of any project organisation structure or any approach to designing organisations is that the success of the structure in achieving its objective can only be measured against the client's satisfaction with the completed project.

The client's expectation at the very beginning of the project is that it will be fully satisfied with the outcome. The components of the client's satisfaction can be taken as function, including aesthetics, price and time, and the client expects to get the project it wants at the price quoted on the date promised. Whilst such criteria are typically used (Bresnen & Haslam 1991), and are initially appealing they are not without difficulties (Ward *et al.* (1991) such as measuring goal attainment, trade-offs of objectives, goal setting and external factors.

The project team's objective will be to provide what the client wants and to achieve this they will have to mitigate and harness environmental forces acting upon the project. The object in designing the organisation structure is to provide one which has the greatest capability to produce the project required within the environmental conditions.

Assuming that the people involved in the project have the requisite skills to carry out their work, the success of the project organisation in dealing with the project is the difference between the client's expectation at the beginning of the project and its satisfaction at completion. The greater the uncertainty present on a project, the greater the opportunity for the achieved outcome to fall short of the client's expectation. If the client is completely satisfied with the project on completion, it can be said that the project organisation structure used was suitable for that project in those conditions and performed satisfactorily. The problem is in measuring any shortfall in client satisfaction adequately so that judgements can be made about the performance of organisations that do not produce projects that fully satisfy clients. This is particularly relevant to the value of comparative studies of different organisational approaches. Ideally, objective quantitative means are required for measuring client satisfaction and environmental forces but these are not yet available.

Nevertheless, it is possible to make an assessment, albeit relatively subjectively, of a client's satisfaction with a project outcome and of the strength of environmental influences (Walker & Wilson 1983). This can be useful in making judgements about the success of project organisations. It is important to remember these issues when powerful claims are made in favour of certain organisation arrangements.

For clarity this discussion has made two major assumptions. One is that the client can be readily identified. The other is that it did not relate the evaluation to the time at which it was made.

The client can be difficult to identify for many projects, particularly those in the public sector. In order to obtain a reliable and comprehensive evaluation it is necessary to consult the whole range of people with a claim to make a judgement on a building. For instance, the project team, developer, occupier, owner, workforce and the general public who may use the building, will all have their own viewpoint, each of which is valid from the position which they adopt but all of which contain large elements of subjectivity. The particular objective of any evaluation needs to be clearly identified and the evaluation designed to satisfy it. For clearly identified private clients, building for their own use, the objective would be to evaluate against the brief for the project. In contrast, for public sector projects it could be that the users believe the brief to have been incorrectly drawn up so the evaluation would need to be against something other than satisfying the brief. The complexity of the issues involved are well illustrated by Liu (1995) who uses an industrial/organisational psychology approach to the evaluation of project outcomes by members of the project team. She identifies the relationship between project success and participant satisfaction in terms of project complexity and commitment using the hypothesis 'the effects of project success on participant satisfaction is dependent on project complexity and the valence of project success is dependent on the level of aspiration of participant satisfaction as well as project complexity'.

An aspect which causes problems with evaluations is that as time passes, people's perceptions of the building change. In the early phases of planning and feasibility, the building exists only as an idea and reactions to it are to an abstract concept which is only defined in very approximate terms. As the concept of the building is developed through drawings and specifications of materials, reactions are likely to be more detailed but still only related to a model of the building, not reality. It is also possible that reactions to the building could be influenced by the construction process itself in terms of the problems and inconveniences it may have generated.

Reactions to a building immediately upon occupation tend to be exaggerated. Expectations are high, the building is in pristine condition, users are unfamiliar with how the building should be used and there are frequently teething troubles with technical aspects. The building then matures, it becomes taken for granted by users, owners, and the general public, it is altered and adapted as demands on it change, it may be rehabilitated but it eventually moves towards the end of its economic life. At each of these stages perceptions of the building will be judged against different criteria by the different classes of people with an interest in it.

This scenario is recounted to impress that, in evaluating the outcome of projects, it is necessary to spell out clearly the perspective being taken (e.g. user, owner, project team) and the criteria against which the evaluation is being carried out. It also illustrates the problems and perhaps, inappropriateness of attempting to arrive at a single measure of the outcome of a project across all perspectives of a building.

Presentation of project organisations

Increasingly, throughout the world, invitations to bid for contracts require bidders to submit details of their proposed organisation structure and approach to the project in addition to the bid price. Quite reasonably, clients wish to be assured that the bidder has a sound and realistic approach to the organisation and successful completion of the project. The problem faced by the bidder, assuming that an acceptable strategy for the project can be produced, is that of convincing the client that the bidder can organise successfully to achieve what is claimed. Propositions that adopt a conventional approach will not necessarily be accepted on face value as many overseas clients will not be familiar with the style and terminology. Frequently, such an approach will not be compatible with the client's needs or the local environment. In many cases, a technique is needed that will explain clearly and convincingly to clients how the project is to be tackled.

Obviously, an LRA will not by itself fully satisfy the client's requirements but it can provide a powerful illustration to the client of how the bidder proposes to organise to carry out the project. It also provides a valuable focal point for the co-ordination of other documentation which will be required to convince the client that the bidder has the capability to complete the work successfully. For example, it is usual for the client to require details of the firms and personnel the bidder intends to involve in the project. Such descriptions can be linked to the LRA by cross-referencing their descriptions to their positions on the LRA. The LRA also follows the style of a network and if it is intended to submit network with the bid, this can also be co-ordinated and cross-referenced to the LRA. Such a package can look very convincing to a client and the client's confidence in the bid can increase, with a consequently greater chance of success.

The large variety of organisation approaches and contractual arrangements which are available and necessary to solve the wide range of different projects on offer throughout the world make it essential for bidders to adopt a creative and constructive method of putting their approaches to clients. The LRA can be used to illustrate any of these approaches because of its flexibility in being able to represent organisations at various levels of detail. It is therefore suitable for bids for construction only, for design-and-build bids and for design-only bids. It is particularly valuable for joint ventures as it will clearly show the aspects of the projects in which the various partners are to be involved. The definitions of the matrix symbols will be most important in such arrangements, as they will define the responsibilities and authority of the various participants.

A particular advantage of the LRA technique in the presentation of proposals is that it should explain to clients what their level of involvement in the project is expected to be. This can enable detailed discussions to be held with the client prior to the bid being submitted, which will be advantageous to the bidder in clarifying the client's requirements in terms of both the project brief and of the client's expectation of involvement in the decision-making process and organisation of the project. If, upon receiving a bid, a client can clearly see not only the design and the price but also how they are to be achieved and what is expected of the client in the process, the client is likely to be assured that the bidder is a professional and competent organisation.

Chapter 13
Case Studies

Introduction

This chapter aims to illustrate some of the ideas which have been discussed earlier from a theoretical basis. Such concepts as differentiation, integration, decision points, authority and power structures, etc. are identified within real projects and their effect on the relative success of the project are identified.

The pattern adopted in each case study is generally to identify:

(a) The primary, key and operational decision points and consequently the project conception, inception and realisation processes.
(b) The extent of the differentiation within the project team and the amount of integration provided to deal with it.
(c) The level of integration of the client. Naturally, the greater the integration of the client the more likely is the success of the project.
(d) The managing system and the operating system and the extent to which they are differentiated. Members of the managing system should not normally be undertaking operating system activities (e.g. designing) as their tasks are differentiated on the basis of skill. A lack of separation of these systems can have an adverse effect on the outcome of the project due to a lack of objectivity on the part of the person undertaking dual roles.
(e) The complexity of the managing system itself. Ideally the managing system should consist of one person from the client organisation and one from the project team (e.g. the project manager). However this ideal is rarely attainable for many reasons, a major one of which is the 'political' structure of the client organisation. The more complex the managing system becomes the more likely it is to have an adverse effect on the outcome of the project.
(f) The pattern of authority within the managing system and the use of power where these influences are seen to have had an effect on the project outcome.

The technique of linear responsibility analysis described in the last chapter was the vehicle for gathering, collating, presenting and analysing the data to identify the features described above.

In order to decide whether these features had affected the overall outcome of the project a measure of the client's satisfaction with the project was established together with details of specific deficiencies the client had identified.

Whilst techniques for satisfactorily quantifying environmental forces do not yet exist, it is important to analyse the results of a project team's performance within the context of the environmental forces acting on the project. Therefore, in each case study the environmental forces acting on the project are described and taken into account in arriving at conclusions.

A more detailed account of the case studies, which includes the research methodology and the results in greater detail, can be found in the publications to which they are referenced. However, on occasions it is necessary to include certain aspects of methodology. In such cases, rather than repeating the general approach taken in each case study, it is only covered in the first one.

A conventionally structured project

The project

This was a relatively small extension to an existing building to provide a butchery facility for a wholesale food company (Walker & Hughes 1986). The contract was for about £0.4 million (at 1996 prices) and, whilst small, it was nevertheless a considerable investment for the client which was a modest family-owned company. Any delay, overrun on costs or unsatisfactory performance could significantly affect the prosperity of the company.

The project was organised and managed on conventional lines. That is, the architect was both designer and manager of the project, other consultants were from separate private practices and the contract was let in competition using bills of quantities as the pricing mechanism. The consultants consisted of a real estate surveyor, quantity surveyor, services engineer and structural engineer. The client had no 'in-house' construction expertise.

The contract was let on the basis of a competitive tender to a medium sized local contractor. Significantly, it was decided that refrigeration work, which was extensive, should be let as a separate contract outside the main building contract.

Project outcome

The client's view of the overall outcome of the project is illustrated in Table 13.1.

In terms of specific deficiencies the client and the project team identified the following:

(1) *Shortfall in quality of the finish to the walls and floors*
For purposes of hygiene the finish to the floors and walls had to be a smooth

Table 13.1   Client satisfaction.

Degree of satisfaction	Function	Time	Price
Very satisfied			
Satisfied	X		
Adequate			X
Dissatisfied		X	
Very dissatisfied			

continuous membrane. When placed it was unsatisfactory and had to be re-ground and repaired. This took a long time to achieve as it was difficult to get the nominated sub-contractor to return to site to complete the work. The floor was installed before the ceiling and the work to the services in the ceiling contributed to damage of the floor.

(2) *Delay in completion of the project*
The client expected to have his building completed in June but in fact it was the following November before it was finally completed, even then defects needed rectification. A one-month delay occurred due to delay in the design stage and four months in the construction stage. The latter was in fact a four-month delay on a seven-month construction programme. Details are as follows:

(a) Delay in the design stage. The client received a final cost estimate five weeks before the date for inviting tenders. The estimate was substantially above what he was prepared to pay for the building. There was a month's delay in the start on site due to the need to propose reductions to the scheme and the time required to negotiate a reduced tender with the lowest tenderer. This reduction was made on the basis of addendum documents. Tenders were called on the original documentation on the assumption that this would prevent further delay.

(b) Delay during construction. There were a number of contributory factors leading to this delay:

  (i) Very severe weather conditions at the beginning of the contract.
  (ii) Delay on the part of the nominated sub-contractor for the roofing and cladding in commencing on site due to an industrial dispute on another project.
  (iii) Due to the addendum to the original design, which was negotiated after tenders were received, many drawings had to be redrafted and were late being issued to the main contractor.
  (iv) The refrigeration contractor (who had a direct contractor with the client which was not part of the building contract) could not commence his work on programme due to delay to the main contract. Nor, when the delay became apparent, was it possible to adequately integrate his work whilst the main contract was being completed.

(3) *The cost of the project was higher than the client's cost limit*
The first estimate for the project was 49 per cent above the client's cost limit. The final estimate before going out to tender was 33 per cent above. The original tender was 31 per cent above the client's cost limit and the client accepted a reduced contract sum on the basis of an amended scheme which was 8 per cent above his original cost limit. The final account cost was 7 per cent above the contract sum.

Environmental forces

(a) The level of uncertainty/certainty in the mind of the client regarding the project at the outset and then during the development of the project is represented by the following:

  (i) A decision had been taken to acquire an existing butchery business which leased existing premises which had been condemned on environmental grounds.
  (ii) A decision had been taken to develop a site under option adjacent to the firm's headquarters to accommodate the butchery business.
  (iii) The opportunity had been taken to extend the lease on the existing butchery premises until the new building was completed but there was pressure to vacate on environmental grounds.
  (iv) The outline technical and functional requirements for the project were known by the client and the cost limit had been decided by him but these two components were not reconciled by the client at the outset.
  (iv) During the process of realisation the decision was taken to accept a reduction in size of the development to meet an acceptable cost which caused a hesitation on the part of the client whether or not to proceed with the project.

(b) The extent of the conflict between participants which existed on the project is represented by:

  (i) A general feeling on the part of the client that the building process was not able to control projects to the same degree as he was able to control his own business.
  (ii) Difficulty in reconciling with the Planning Authority the need to provide a canopy which would have had a significant effect on the cost of the building.

(c) The relative complexity of the project determined by external expectations and requirements of the process to be undertaken in the completed building are represented by:

  (i) An adequate site area was available.
  (ii) The location on the site was pre-determined.

(iii) The areas required for different usages were known and easily accommodated.
(iv) A conventional structural solution was possible and was adopted.
(v) Extensive refrigeration was required.
(vi) Extensive hygiene facilities requiring a high standard of finishes were needed.
(vii) The project had to match existing simple elevations and functional satisfaction was dominant.

Primary and key decisions

The two primary decisions:

(a) to acquire an existing butchery business and
(b) to provide a new building for it at the existing head office site.

were taken practically simultaneously. The acquisition was part of the client organisation's expansion plans. These decisions were identified but as the systems which led to them took place within the client's organisation it was not possible to analyse them in detail.

The project realisation process was clearly identified and analysed. The primary decisions and the key decisions within the project realisation process are listed in Table 13.2. In order that decisions can be taken and the project progressed, it is obviously necessary for certain tasks to have been performed. Their relationship to the key decisions is also shown. Seven main sub-systems are identified. They are separated by 'pinch points' created by primary and key decisions. Operational decisions occur between the tasks and are listed at the right hand side of the table but in order to avoid too fine a level of detail they are not considered further.

Table 13.2 Decisions and sub-systems.

Decisions	Task sub-systems
Start point	*System A Project conception*

Primary decision no. 1
Acquire an existing
butchery business
	System B Project inception
	1st main sub-system
	  1. Acquire existing butchery business
	  2. Negotiate extension of lease
	  3. Identify outline building requirements

Primary decision no. 2
Decision to construct new
building on a site on which
an option was held, outline
requirements defined
	System C Project realisation
	2nd main sub-system
	  4. Preliminary programme
	  5. Contractual arrangements
	  6. Development of outline building proposals

Table 13.2   (continued).

Decisions	Task sub-systems
	7. Development of outline refrigeration and services proposals
	8. Preliminary structural proposals
Key decision no. 1 Agreement to proceed on basis of sketch design	3rd main sub-system 9. First estimate 10. Detailed spatial design 11. Architectural detailed technical design 12. Refrigeration detailed design 13. Structural detailed design 14. Services detailed design 15. Second estimate
Key decision no. 2 Estimate too high, scale of acceptable reductions indicated, proceed to tender on existing scheme subject to reductions	4th main sub-system 16. Prepare building contract documentation 17. Prepare first addendum to scheme design 18. Prepare second addendum to scheme design 19. Negotiate reduction with lowest tenderer 20. Main contract tender action
Key decision no. 3 Refrigeration to be a separate contract	5th main sub-system 21. Prepare refrigeration contract documentation 22. Refrigeration contract tender action
Key decision no. 4 Acceptance of tender for main contract subject to reduction	6th main sub-system 23. Construction programme 24. Construction (main contract) 25. Construction (nominated subcontractors) 26. Construction (domestic subcontractors) 27. Supporting spatial material 28. Supporting technical architectural material 29. Supporting technical services material 30. Supporting technical structural material 31. Cost statements 32. Final account 33. Certificate of final completion
Key decision no. 5 Acceptance of refrigeration tender	7th main sub-system 34. Refrigeration installation 35. Supporting technical refrigeration material 36. Approval of refrigeration work 37. Settlement of refrigeration account
Key decision no. 6 Contractual obligations discharged	

Note: Tasks and sub-systems are not necessarily sequential as shown. In many cases they are concurrent, in particular sub-systems 4 and 5 were parallel as were sub-systems 6 and 7.

Differentiation and integration of the operating system

From the linear responsibility analysis it was possible to identify the amount of differentiation between the members of the project team on the basis of the features described in Chapter 7. It was also possible to quantify the level of integration provided by the architect who was managing the project.

A high level of differentiation was present in the system most of which was of the most complex type – that is, due to technology, territory, time and sentience relating to the company by which the person was employed and to profession. It is also clear that the amount of integration provided was extremely low.

Integration of the client

On this project the two primary integrators identified were:

(a) the managing director of the client organisation
(b) the project architect.

Two measures of integration were used – good and partial. Good integration refers to the presence on both integrators in the control loops of the linear responsibility analysis. When partial integration occurs one of the integrators appears in a relationship other than in a control loop, e.g. consultation.

The level of integration of the client was at a low level, particularly between tasks. Of that which occurred, the majority was only partial.

The differentiation of the managing and operating systems

The architect who was managing the project also acted operationally for 30 per cent of the tasks involved in the project. In each of these situations there was a potential lack of objectivity on the part of the architect as he was both designing and managing himself and others at the same time.

The complexity of the managing system.

In only about 25 per cent of cases did the managing system have two managers, i.e. one from the client and one from the project team as recommended as being ideal. On just under one third of cases the managing system only had one member and for the remainder no one was managing.

Continuity of the managing system is also important and it was found that there was discontinuity in the major integrating activity of boundary control between over 90 per cent of tasks and that at key decision points discontinuity occurred on every occasion.

Authority and power

The authority pattern on the project was that which is implicit in conventionally organised projects. Whilst not formally documented, the architect had formal authority over the project team and no member of the project team had sufficient informal authority to countermand the formal authority of the architect.

As the client had no experience of construction, the architect had informal authority over the client. There were no 'power plays' on this project as all authority was employed legitimately.

Conclusions

The process was divided into systems by primary decision points and into the main sub-systems by key decision points.

However the quality of key decision no. 1, which did not clearly define the client's cost limit, meant that feedback did not operate effectively and so reduced the effectiveness of subsequent sub-systems.

The analysis showed that:

(a) Whilst the process was heavily differentiated (i.e. the large majority of differentiated relationships were of the most complex type) the level of integration provided was erratic and generally low.

(b) The managing and operating systems were mainly undifferentiated in the first two sub-systems but differentiation improved in the other sub-systems.

(c) The constitution of the managing system was erratic so that membership was not consistent. There was very high discontinuity of the managing system between tasks, particularly at key decision points where it was 100 per cent.

(d) Integration of the client and the project team was erratic. It did not occur in the first system where the client did not involve the project team. It then improved substantially in the next three systems and sub-systems only to tail off again except for the construction sub-system where it was partial.

(e) The architect as designer and manager of the project with implied formal authority did not exercise his authority sufficiently to ensure that the systems operated effectively.

The problems focused on key decision no. 1 at the end of the second sub-system and the lack of definition of the client's cost limit for the project. This occurred through a shortfall of integration within the sub-system and between the client and the project team. The former was due to a deficiency in establishing reciprocal interdependencies between tasks and thus a lack of appropriate consultation between contributors. This followed through to deficiency in the integration of the client between tasks. As a result the client did not meet the quantity surveyor at this stage and a cost limit for the project was not clearly established and understood by all the contributors.

The process undertaken in the third sub-system was therefore not in accordance with the client's requirements and, because the same relationships occurred in the third sub-system as occurred in the second the deviation was not corrected. The feedback mechanism was therefore faulty as development of the project was being measured against a goal which had been inappropriately defined.

The deviation was discovered at the end of the third sub-system and the fourth sub-system had to correct it by reducing the scope of the project. The fourth sub-system was structured exactly as the previous two and was not able to achieve an appropriate correction in terms of cost but the client accepted a higher estimated cost for the project. In addition, the need for this additional sub-system induced a delay which meant that the client's requirement in terms of time for completion of the project could not be met. The delay which was induced was not only in design time. There was a 'knock on' effect into the sixth sub-system (construction) through delay in issuing drawings to the contractor as they had to be redrafted following amendment to the design.

Running parallel with these sub-systems were the fifth and seventh sub-systems which concerned a direct contract with the client for refrigeration. The lack of integration of these sub-systems with the main design and construction sub-systems meant that delay of the building contract was likely to delay the refrigeration contract which in fact was the case.

Essentially the major causes of problems on this project were a poorly integrated operating system, and poor integration of the client with the project team. This overall shortfall of integration produced a deviation from the client's requirements. It could not be fully corrected because it was not discovered until a late stage in the process when efforts were made to mitigate its effects. This resulted in the most significant of the project's deficiencies, namely delay in completion and a price which was over the budget. In addition, members of the project team were not involved by the client on the first system of activity. It is probable that, had the project team been involved at this stage, the deficiencies would not have occurred. Not only may all members of the team have been more aware of the client's cost limit but their presence may have reduced the duration of this system so as to reduce the pressure of time on subsequent sub-systems.

The sixth sub-system (construction) was typical of those undertaken as a result of a conventional competitive tender under the JCT standard form of contract. The contractor acted in an integrating role for the work for which he was responsible and the architect adopted an approval/recommendation role strictly in accordance with the contract.

Apart from the delay discussed previously, the only deficiency which occurred in this sub-system was the lack of satisfaction with the floor and wall finishes and the need for subsequent remedial work. This contractual arrangement creates a complex managing system which could have contributed to a lack of awareness of the high quality of finish required.

Environmental influences

There was a low level of influence of environmental forces on the client's organisation and hence the process of providing the building was relatively stable in this respect. The action of environmental forces directly on the building process was more pronounced and occurred in the sixth sub-system (construction).

The environmental forces which influenced the decision to submit the project to competitive tender created the complex management system in the construction sub-system which contributed to the problem of the wall and floor finishes.

The other environmental influences were the very severe weather and the delay in the cladding sub-contractors arriving on site due to a strike elsewhere. It is unlikely that any alternative managing system could have overcome these events to any greater extent on such a short duration contract.

Attributes

It is important to note, in spite of the difficulties encountered in meeting cost and time objectives, the completed building was both spatially and technically satisfactory to the client and eventually satisfactory in terms of quality once the wall and floor finishes had been rectified. Any degree of implied criticism of the performance of the process should be judged against this success achieved in the project outcome.

A project management structured project

The project

This project was an extension to an existing pharmaceutical manufacturing plant to provide a high-rise warehouse, services block and packing line (Walker & Hughes 1984). The contract was for about £4.5 million (at 1996 prices). The client was the British manufacturing plant of an American company.

A professional engineering consultancy firm with 'in-house' expertise in structural and services engineering acted as professional advisors. In addition to providing these professional skills they also provided the project management on behalf of the client. They engaged an architect for the architectural aspects only. A quantity surveying firm was engaged directly by the client to work with the other professions.

For reasons which will become clear, the contract was let in two parts, both by competitive tender. Contract no. 1 was for all work up to and including the sub-base to the ground floor slabs, including piling. Contract no. 2 was let to a different contractor for the superstructure and all other work.

Contractor no. 1 was a medium-sized firm specialising in sub-structure work and earthmoving, and contractor no. 2 was a medium to large national contractor – a public company which engages in contracts over a wide range of building and civil engineering.

Project outcome

The client's view of the overall outcome of the project is illustrated in Table 13.3.

Table 13.3   Client satisfaction.

Degree of satisfaction	Function	Time	Price
Very satisfied			X
Satisfied	X		
Adequate		X	
Dissatisfied			
Very dissatisfied			

In terms of specific deficiencies, the client and the project team identified the following:

(1) *Low quality of the warehouse floor*
The warehouse facility utilises forklift trucks for stacking pallets. These trucks have an unusually high reach, and consequently the tolerance in the level of the finished floor is critical to their smooth operation. The problem was partly due to the masts on the trucks being out of plumb, nevertheless the trucks still rocked due to the uneven floor after the masts had been corrected. This had the effect of slowing the trucks down so they could not achieve the expected number of pallet stackings per hour. The need to rectify this problem caused a delay in the construction stage.

(2) *Dissatisfaction with the manufacturing area*
One of the major reasons for embarking on the project had been the need to create more manufacturing space. This was to be done by converting part of the existing low-rise warehouse into manufacturing area after transferring part of the warehousing capacity to the new high-rise warehouse. The space available for conversion was unsatisfactory. Conversion of the extra manufacturing area generated did not form part of this project. It was seen by the client as a bonus and detailed appraisal of alternative solutions was not required to be considered by the project team as part of the total problem. This was not a shortfall of this project as such, although it was a source of the client's dissatisfaction with his total concept, as the two projects were not fully considered as a whole.

(3) *Delay in completion of the project*
The client expected to have his project completed in April but it was the following December before practical completion.

(a) Delay in the design stage. The project had a brief in the form of a comprehensive document of detailed functional and technical requirements and cost and time plans. The client's organisation had to have this document approved by head office in the USA before they could proceed. This approval took considerably longer (three months) to gain than was originally envisaged. The delay in approval of the brief document created a situation where consultant's staff had been allocated to other projects, as well as the need to let the project in two parts. The demands of this project then conflicted with the demands of other projects being worked on by the consultants.

(b) Delay in the construction stage. As mentioned above, the problem of providing a level floor in the warehouse delayed the construction of the project; further to this there was a shortage of bricklayers on the site due to the fact that the main contractor was worried about employing bricklayers who seemed to be closely involved in an industrial dispute on a nearby site. Also, the decision to award the contract in two stages, which was supposed to claw back some time, did in fact cause further delay to the programme proposed for the construction of the superstructure.

Environmental forces

(a) The level of uncertainty/certainty in the mind of the client regarding the project at the outset and then during the development of the project is represented by the following:

 (i) The functional and technical requirements were known at the outset.
 (ii) The project formed part of a planned expansion programme.
 (iii) The head office and site personnel had a commitment to the project.
 (iv) During the process of realisation the programme was compressed for financial and budget reasons.
 (v) The requirement for early completion eased during construction due to client's manning levels and reduction in demand.

(b) The extent of the conflict between participants which existed on the project is represented by:

 (i) There was some conflict within the architectural practice about the type of work to be undertaken.
 (ii) A change in services engineering manager towards the end of the third sub-system of activity without a replacement until the fourth sub-system of activity.
 (iii) A change in the main contractors' site agent without a fully satisfactory replacement.
 (iv) A shortage of bricklayers.

(c) The relative complexity of the project determined by external expectations and requirement of the process to be undertaken in the completed building are represented by:

(i) The location of the site was predetermined.
(ii) Areas of different usage were known.
(iii) There were tight constraints on the location and relationship of areas.
(iv) A high level of specialist equipment had to be housed.
(v) The relationships to existing facilities were important.
(vi) The warehouse structure was complex and a variety of structures were used.
(vii) There was a complex services requirement for the specialist equipment and dust extraction and ventilation and temperature controls were important.
(viii) Warehouse racking and trucks were at the extremes of technology.
(ix) The project had to match simple existing elevations.

Primary and key decisions

The two primary decisions, that is the plan to expand the plant and the decision to provide for expansion through the construction of new buildings were taken about twelve years previously as part of the company's strategic planning and they were confirmed prior to embarking on the project. It was therefore not possible to analyse the project conception and inception process as they had taken place so many years previously.

The project realisation process was clearly identified and analysed. The key decisions within the project realisation process are listed in Table 13.4. In order that decisions can be taken and the project progressed, it is obviously necessary for certain tasks to have been performed. Their relationship to the key decisions is also shown. Four main sub-systems are identified, they are separated by 'pinch points' created by key decisions. Operational decisions are listed down the right hand side of the table.

Differentiation and integration of the operating system

A large part of differentiation between tasks was of the most complex type. This was also true within tasks but to a lesser degree. It was also clear that a high level of integration was provided to cope with the extensive differentiation.

Integration of the client

On this project the two primary integrators identified were:

(a) The client's project engineer
(b) The project manager.

The level of client integration was lower than it should have been, at just less than 50 per cent evenly split between good and partial integration.

Table 13.4 Decisions and sub-systems.

Decisions	Task sub-systems
Primary decision no. 2 Decision to achieve the client's purpose through the development of a new building	*System C Project realisation* 1st main sub-system 1. Identify need for project 2. Define outline requirements 3. Establish budget estimate 4. Present for budget inclusion
Key decision no. 1 Approval to proceed with feasibility work given by parent company	2nd main sub-system 5. Programme proposals 6. Contractual proposals 7. Spatial proposals 8. Technical structural proposals 9. Technical services proposals 10. Technical architectural proposals. 11. Financial proposals 12. Consolidate brief document 13. Capital expenditure presentation
Key decision no. 2 Permission by parent company to proceed with detailed design. Commitment to full design fees, based on approval of brief.	3rd main sub-system 14. Programme details. 15. Working drawings 16. Technical structural details 17. Technical services details 18. Technical architectural details 19. Contract details 20. Contract documentation 21. Tender presentation
Key decision no. 3 Acceptance of tender. Agreement to construct. Commitment to construction costs and contractual obligations	4th main sub-system 22. Substructure programme preparation 23. Substructure construction (contract no. 1) 24. Substructure construction (nom. subcontract) 25. Final account contract no. 1 26. Certificate of completion contract no. 1 27. Superstructure programme preparation 28. Superstructure construction (contract no. 2) 29. Superstructure construction (nom. subcontract) 30. Superstructure construction (dom. subcontract) 31. Supporting spatial material 32. Supporting technical structural material. 33. Supporting technical services material 34. Supporting technical architectural material 35. Cost statements 36. Final account contract no. 2 37. Certificate of completion contract no. 2
Key decision no. 4 Certificate of final completion. Contractual obligations discharged	

Note: Tasks are not necessarily sequential as shown. In many cases they are concurrent.

Differentiation of the managing and operating systems

The project manager acted in an operational as well as a project management role for 24 per cent of tasks. In each case there was potential for a lack of objectivity on the part of the project manager.

The complexity of the managing system

The complexity of the managing system was such that there was normally either three or four managing positions per task. This number is greater than the ideal of two managing positions.

On the face of it, continuity of the managing system between tasks was good as discontinuity, occurred for only 5 per cent of tasks. However, this view would be deceptive as this discontinuity occurred mainly at key decision points, resulting in discontinuity at 75 per cent of key decision points.

Authority and power

The 'in-house' project engineer had formal and informal authority over the project team, including the project manager. This was particularly manifest in the detailed brief for the project which laid the basis for its successful functional outcome. However the project engineer did not have formal authority over the local directors who in turn did not have formal authority over the head office.

The project manager had formal and informal authority over the project team. His informal authority was particularly pronounced over the engineers in the practice to which the project manager also belonged.

There were no 'power plays' on the project.

Conclusions

The clauses of the main deficiencies of the complete project can be summarised generally as:

(a) Lack of integration of the client and the building process
(b) Lack of integration between sub-systems
(c) Complexity of the managing system.

Although sound integrating mechanisms were generally provided, nevertheless a shortfall of integration occurred through the lack of the ability of the client to respond to the integration demands placed upon him in the 1st main sub-system. The whole of this sub-system took place in the client's organisation with no reference to the project team. The authority of the local board of directors over the project engineer meant that his advice to incorporate the project team at this time was not acted upon. As a result this project and the conversion of the vacated area to manufacturing was not co-ordinated leading to dissatisfaction within the combined outcome.

Integration also broke down at the boundary of main sub-systems 2 and 3 (identifiable from the LRA) due to the client's lack of response which induced a delay at a key decision point due to the delay in the approval by the client's head office. Again the authority of the head office over the local directors meant that the latter's advice was not acted upon. This in turn generated the decision to award the building contract in two stages which created delay in main sub-system 4. The delay also contributed to conflict within the consultants' offices due to the demands of other projects which then caused further delay in sub-system 3.

In addition, lack of integration between sub-systems 3 and 4 caused by the award of the building contract by competition, and the consequential conditions of contract, prescribed the managing system of sub-system 4. This resulted in the introduction of new managing units from the contracting firms creating a complex managing system with duplication and differentiation of managing roles. This contributed to lack of awareness of the tight tolerances needed in the warehouse floor construction and the consequential delay in rectifying it. In addition delay was also caused in sub-system 4 by awarding the contract in two stages without consultation with the contractor's element of the managing system leading to insufficient allowance being made in the programme for the super-structure contractor to set up his operation.

Environmental influences

There was a low level of influence of environmental forces on the client's organisation and hence the process of providing the building was relatively stable in this respect. The action of environmental forces directly upon the building process was more pronounced and, with the exception of the resignation of the Services Engineering Manager, all manifest in sub-system 4. The most significant were those which influenced the decision to submit the project to competitive tender referred to above.

The other major environmental influence which was not finally overcome was the shortage of bricklayers which contributed to delay. Other environmental influences were mitigated without detriment to the programme.

Attributes

There was a high level of satisfaction with the completed project and this is reflected by the way in which the organisation structure corresponded to the model's propositions. The pattern of decision points could be clearly identified and whilst there were specific shortfalls of integration, the general level was high. The managing and operating system were clearly differentiated but the differentiation of the managing system itself was erratic as described, as was the integration of the client.

Any degree of implied criticism of the performance of the process should be judged against the high measure of success achieved in the project outcome. It is interesting to observe that the outcome deficiencies were generally as a result of organisational defects within the managing system itself rather than in the way in which the operating system was managed or within the operating system itself.

A multi-disciplinary practice project

The project

This project was an engineering factory and offices with a contract sum of £2 million (at 1996 prices) (Walker & Hughes 1987). The client was an engineering company which is part of a privately owned group of companies. The group is involved in a wide range of engineering design and manufacture in terms of both scale of project and technological content. The engineering company for whom the project was constructed reflects the diverse activities of the group.

A multi-disciplinary industrial and commercial property consultancy firm with in-house expertise in general practice surveying, architecture, structural design, services and quantity surveying acted as project managers and provided all professional services.

For reasons which will become clear, the contract for construction was let in two parts. The first part, for the factory, was let by competitive tender. The second part, for the offices, was negotiated with the contractor appointed for the factory. The contractor was a medium sized national contracting firm which is part of a privately owned group including a property development company and a private house building company.

Project outcome

The client's view of the overall outcome of the project is illustrated in Table 13.5.

Table 13.5   Client satisfaction.

Degree of satisfaction	Function	Time	Price
Very satisfied			X
Satisfied	X		
Adequate		X	
Dissatisfied			
Very dissatisfied			

In terms of specific deficiencies, the client and the project team identified the following:

(i) *Function*
The client felt that the offices might prove to be too small, that too many toilets had been provided and not enough locker space. The client was not fully satisfied with the quality of the construction work, particularly bricklayer's work.

(2) *Delay in completion of the project*
The client expected to have his building completed by July but in fact it was the following November before the factory was completed and the following May when the offices were completed with the exception of defect rectification. The construction programme dates were October for the factory and the following February for the offices. There was therefore delay during the design stage and some slippage in the construction period.

(a) *Delay in the design stage.* There was difficulty in gaining the client's agreement to the final design. As a result, it was agreed to proceed with the construction of the factory as soon as the design was approved and to continue work on the design of the offices. The price for the office construction was negotiated with the successful bidder for the factory. Nevertheless, this lack of agreement by the client created delay in the design stage which the splitting of the project into the two phases could not fully overcome. In addition, the client believed that the consultancy firm was over-committed to other projects during the design phase; the consultants did not agree.

(b) *Delay during construction.* This delay to the completion beyond the date in the construction programme was due to additional foundation work being required due to a pond on the site requiring more extensive work than originally anticipated. There was also particularly inclement weather in the winter during which construction took place.

Environmental forces

(a) The level of uncertainty/certainty in the mind of the client regarding the project at the outset and then during the development of the project is represented by the following:

    (i) There was a clear recognition of the need for alternative premises.
    (ii) It was assumed that alternative premises could be provided by an existing building to be purchased or leased. There was a reasonably clear idea of the outline functional requirement of the factory.
    (iii) There was continuous hesitation regarding the desirability of pursuing the project due to the client's concern over the economic/political situation, variation in distribution of demand across the company's range of production, and labour problems. This hesitation was

compounded by the fact that this project represented the largest capital commitment ever entered into by this company.

(iv) There was a change from the intention to provide the facility by acquisition of an existing building to its provision by constructing a new building.

(v) There were pressures on the client due to fluctuating prospective orders from a large US customer.

(b) The extent of the conflict between participants which existed on the project is represented by:

(i) The client believed that the delay to the completion of the offices and the need to split the contract into two phases was due to an over-commitment to other projects by the consultants. The consultants believed it was due to the client's inability to make decisions about his requirements.

(ii) The client believed that his standards and values arising from his company's work particularly in relation to workmanship, were different from those of the consultants and the construction industry.

(iii) The site agent for the factory construction resigned and it was felt by the client that he was not replaced by a man of equivalent calibre for the office construction.

(iv) There was a shortage of bricklayers who were prepared to lay the reconstructed blocks manufactured by one of the client's subsidiary companies.

(c) The relative complexity of the project determined by the external expectations and requirements of the process to be undertaken in the completed building are represented by:

(i) There was adequate site area with no severe constraints on location.
(ii) The floor areas required were not predetermined.
(iii) The residual value was important.
(iv) Flexibility was required to a degree which was technically difficult.
(v) Functional requirements were dominant.
(vi) Noise and vibration were not to be transmitted to the offices from the factory.
(vii) Site problems were encountered (pond volume more extensive than anticipated). Straightforward services were required.
(viii) The aesthetic demands were low. Functional efficiency was to be provided at the expense of aesthetics.

Primary and key decisions

The decision to acquire real property was taken within the client organisation and it was therefore not possible to analyse the project conception process. The project inception process which resulted in the decision to construct a new building was analysed and documented and includes an examination of the possibility of acquiring an existing building before the decision to construct a new building was taken. The project realisation process was clearly identified.

The primary and key decisions within the project inception and realisation systems are listed in Table 13.6. In order that decisions can be taken and the project progressed, it is obviously necessary for certain tasks to be undertaken. Their order and relationship to the decisions is also shown. Four main sub-systems are identified, one of which has three further sub-systems, making seven sub-systems in total, each being punctuated by the 'pinch points' created by primary and key decisions. Operational decisions are listed down the right hand side of the table.

Table 13.6 Decisions and task sub-systems.

Decisions	Task sub-systems
Start pont	*System A Project conception*
Primary decision no. 1 Decision to achieve the client's purpose through the acquisition of real property	*System B Project inception* 1st main sub-system 1. Confirm need for project 2. Identify outline requirements 3. Identify existing buildings 4. Feasibility studies 5. Identify site 6. Feasibility studies 7. Site negotiation
Primary decision no. 2 Decision to lease site and construct new building	*System C Project realisation* 2nd main sub-system 8. Preliminary programme 9. Develop outline proposals 10. Financial advice 11. Detailed spatial design 12. Architectural and services detailed technical design 13. Structural detailed design 14. Financial control 15. Revised programme
Key decision no. 1 Agreement to proceed to tender on factory but not on offices; office design to be revised and tender negotiated	3rd main sub-system 16. Programme 17. Contractual arrangements 18. Contract documentation (factory) 19. Tender action (factory)
Key decision no. 2 Acceptance of tender for factory; agreement to construct; commitment to construction costs and contractual obligations	3(a) sub-system 20. Detailed spatial design (offices) 21. Architectural and services detailed technical design (offices) 22. Structural detailed design (offices) 23. Financial control (offices)

Table 13.6   (continued).

Decisions	Task sub-systems
Key decision no. 3 Agreement to design of offices	3(b) sub-system 24. Contract documentation (offices) 25. Negotiation of contract sum (offices)
Key decision no. 4 Decision to redesign offices	3(c) sub-system 26. Spatial redesign (offices) 27. Architectural and services technical redesign (offices) 28. Structural redesign (offices) 29. Financial control
Key decision no. 5 Acceptance of office design, price to be established on basis of established on basis of previous documentation and negotiated price	4th main sub-system 30. Construction programme 31. Construction (main contractor's work) 32. Construction (main sub-contractor's work) 33. Construction (domestic sub-contractor's work) 34. Supporting spatial material 35. Supporting technical architectural and services material 36. Supporting technical structural material 37. Cost statements and cost control 38. Final account 39. Certificate of final completion
Key decision no. 6 Contractual obligations discharged	

Note: Tasks and sub-systems are not necessarily sequential as shown; in many cases they are concurrent.

Differentiation and integration of the operating system

Unlike the previous projects, differentiation between tasks is not predominantly of the most complex type. There is also less of the most complex type of differentiation within tasks than for the previous case studies. This situation is brought about by the use of a multi-disciplinary practice. A high level of integration was provided.

Integration of the client

On this project the two primary integrators identified were:

(a) the group chairman of the client company
(b) the project manager.

The level of client integration was higher than for the previous case studies, running at over 60 per cent with the largest proportion of good integration.

Differentiation of the managing and operating systems

The project manager acted in an operational as well as a project management role for 20 per cent of tasks. In each case there was potential for a lack of objectivity on the part of the project manager.

The complexity of the managing system

Generally there were two managing positions per task which is the ideal number. Nevertheless there was still appreciable variation from the ideal.

Continuity of the managing system between tasks was ideal as no discontinuity occurred.

Authority and power

The group chairman of the client company maintained formal authority over the project manager and the project team. The project manager had no informal authority over the group chairman due to the latter's scepticism about the construction process.

The project manager had formal and informal authority over the project team due mainly to him being a partner in the multidisciplinary practice from which the project team came.

There were no 'power plays' on the project.

Conclusions

The process was divided into systems by primary decision points and into sub-systems by key decision points. However, the structure of the sub-systems was strongly influenced by environmental forces acting on the project. These influences generated uncertainty in the client with the result that decisions were difficult to obtain, particularly as the group chairman maintained formal authority over the project, hence the structure of the sub-systems was erratic. The differentiation of the operating system was not of the most complex kind due to the use of a multi-disciplinary consultancy firm, yet the level of integration was extremely high. The level of integration of the client was also high.

Differentiation of the managing system within tasks was generally normal. There was no differentiation of the managing system between tasks. It is particularly important that there is no discontinuity at decision points and this was the case. The level of differentiation of the managing and operating systems was again high. However, in the earlier sub-systems there was some lack of separation.

The organisation of the project subscribed to the propositions of the model to a very high degree. It is therefore to be expected that a high level of client satisfaction would be found with the outcome of the project.

Whilst a high level of client satisfaction was obtained (see Table 13.5), in terms of function the client was nevertheless only 'satisfied' as opposed to being very satisfied and in terms of time his level of satisfaction was only 'adequate'. It is necessary therefore to examine the cause of these shortfalls in the client's total satisfaction. The model proposes that if its propositions are fully met, then the organisation should have the capacity to overcome the environmental forces acting on the project. It is therefore necessary to examine the impact of these forces of the project.

Environmental influences

As stated earlier, there was a high level of environmental forces acting upon the client's firm and hence on the project. They created instability in the process of providing the building which the managing system sought to control. The action of the environmental forces was manifest in the client's hesitation regarding the desirability of pursuing the project which created difficulty in defining the client's requirements and hence created a high level of uncertainty. This situation was compounded by further environmental forces which induced conflict between the client and the project team. This conflict arose through the client's expectation of the standard of project with which he should be provided and his view that the members of the construction industry did not have similar standards. Hence the client did not delegate authority to the project manager nor did the project manager have informal authority over the client.

These factors provided the context within which the project was pursued and were evident in the project organisation's configuration through the lack of positive and progressive key decisions. It was therefore necessary for feedback to be to the client's requirements established at the very start of the process without further refinement throughout the process. Thus the uncertainty surrounding the project created a structure with few key decisions and a fragmentation (sub-systems 3a, b and c) of the decision-making process.

The deficiencies referred to previously were traceable to environmental forces. The functional deficiencies were due to uncertainty of what the client required as reflected by the view that the offices might prove to be too small. In the case of the toilet provision there was a misunderstanding between the client and the project team; the misunderstanding was not discovered due to the hasty decision to proceed to tender for the factory in advance of the offices which was an attempt to overcome the uncertainty surrounding the project. This situation was compounded by a decision during construction to provide a toilet block large enough to accommodate a future extension to the factory which further reflected the uncertainty surrounding the project. The client's perception of a shortfall in the quality of construction work arose from the conflict referred to earlier.

In terms of the delay in completing the project relative to the original expectation, this was again due to the uncertainty generated by environmental forces which meant that final acceptance of the design, particularly the offices, was delayed and redesigning was required. The delay during construction was due to the unexpected need for additional foundation work and also inclement weather which can be seen as typical hazards on construction work.

Attributes

There was a high level of satisfaction with the completed project. However, whilst there was a high correlation of the project organisation's configuration and the model's proposition, the project team was still unable to overcome entirely all of the very strong environmental forces acting on the project. It appears, on the basis of this case study, that even when project teams are structured to cope with environmental forces, there are limits beyond which they are unable to mitigate all such forces. The project team was not involved in the conception stage of the project. It can be speculated that had it been, it may have been better placed to overcome the environmental forces to a greater extent.

Any implied criticism of the performance of the process should be mitigated by the high measure of success achieved in the project outcome in difficult circumstances.

A Hong Kong project with a novel structure

The project

The fourth case study is distinctly different from the proceeding three case studies on many grounds. It is not in the UK, it is a public sector project, it involved the private sector in a novel organisational structure, it was a massive project (about £400 million at 1996 prices). Nevertheless, the analytical techniques applied as a result of the theoretical base developed in this book gave valuable insights into the management of the project (Walker & Kalinowski 1994).

In 1966 the Hong Kong Government decided to establish the Hong Kong Trade Development Council (TDC) and decided simultaneously that a Convention and Exhibition Centre (CEC) would be necessary to support the TDC's activities. From 1966 a number of options for providing a CEC were explored, which included using existing buildings and the possibility of a new building, without reaching a decision. As a result in the late 1970s, the government commissioned the UK-based Earls Court to prepare a report on the feasibility of developing a CEC in Hong Kong. The report recommended that a CEC was feasible. Subsequently, the government decided that a non-statutory board should be established independently of the government to plan and develop a CEC.

The non-statutory board was appointed by the government in January 1983 and was chaired by the then Chairperson of the Hong Kong and Shanghai Bank and included members of the Executive Council (Cabinet) of the government.

The board appointed a firm of exhibition centre specialists to prepare proposals for a CEC. By mid-1983, their proposal, which recommended a site on reclaimed land on the Wanchai waterfront, was accepted by the non-statutory board which recommended that the government develop the CEC on this site.

However, on receiving the recommendation, the government had grave doubts on financial grounds as the economy was weakening at the time. The government's reaction created extreme dissatisfaction on the part of the non-statutory board, the TDC, the business community and even some government officials. Private lobbying took place which generated a significant change in the direction of the project, resulting in the government offering the Wanchai site to the TDC free of charge on condition that the TDC, rather than the government, developed the CEC on the site at no further cost to the government. The TDC accepted the offer. They appointed C-Fin as their professional advisors who later become project leaders.

The project was to be let as a design-and-build contract on a fast-track basis at no cost to the TDC. The reason why this was possible was that the successful bidder would be granted the space above and around the convention and exhibition accommodation to develop, within constraints, what they decided was necessary in order to make the project viable. That is, it had to provide an acceptable return whilst covering the cost of the CEC. In the event the successful bidder decided to build two world class hotels, an office tower and a serviced apartment tower. Some of the office space was allocated to the TDC.

C-Fin's initial task was to develop a brief and an indicative design for the purpose of inviting bids. They appointed Hong Kong architects and services engineers as their consultants for this task.

TDC awarded the contract to New World Development (NWD) in late December 1984 based on an indicative design as developed in outline in NWD. The definition of what was to be provided was so vague at this stage as to contain a clause which stated that: 'NWD shall provide for TDC a first class (world standard) exhibition centre'. The contract for the whole of the design and construction of the CEC was between the TDC and NWD at no cost to the TDC. NWD used Polytown (a project management company set up specifically for the project and wholly owned by NWD) to manage the project. The main contractor was NWD's in-house contractor. C-Fin continued as project leaders commissioned directly by the TDC. The basic organisation structure is shown in Fig. 13.1.

The project outcome

It is generally accepted that all the clients were very satisfied with the outcome in terms of function, cost and time. In terms of function the CEC is regarded as one of the best in the world and the two hotels are successful as are the office tower and service apartments.

In terms of cost, all parties appear to be very satisfied. The TDC have a world

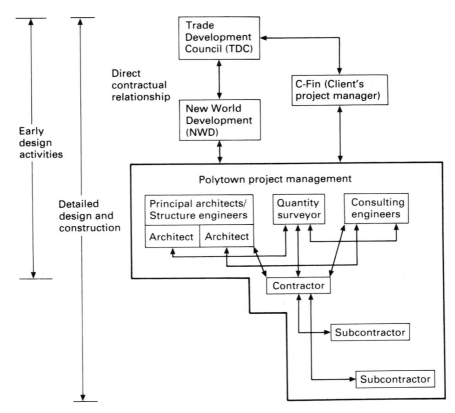

Fig. 13.1 Organisation structure.

class CEC at no capital cost to them (but at an opportunity cost to the government). NWD have not disclosed their financial position on the project but later sections of this paper will explain their position in more detail to justify their satisfaction with the deal.

It is in terms of time that the performance of the project team was remarkable. Although it took 18 years from the establishment of the TDC to the award of the contract for the project, it then took less than four years for the design and construction of the project and prior to that only two years to work up the basis of the contract from what was practically a start-up position.

Environmental forces

The size and complexity of this project means that a more sophisticated method of describing the environment is necessary. Hughes's (1989) micro-environment categories are used as follows:

(1) *Legal/institutional:* No formal contract was used between developer/contractor. Many consultants had worked with the developers before. Familiar conditions of engagement were used. Professional framework was standard practice.

(2) *Technological:* Sheer size made it complex. Some structural elements were extremely complex. Novel approaches to the design were adopted. The scale of the structure can be envisaged from the need for 1200×1200 mm truss girders spanning 48 m and beams 2500 mm deep and a total quantity of steel exceeding 25 000 tonnes all of which were records for a commercial building in Hong Kong. The services provision was extremely complex: acoustic separation, unique sea-water pump house, super-large variable air volume and heat recovery air handling system, innovative smoke control system, world's largest hydraulic lift system, state-of-the-art integrated information and teleconference system, intelligent integrated fire detection, security, public address and building management system.

(3) *Financial:* A novel solution to the economic viability was established.

(4) *Aesthetic:* The aesthetic demands were high. The project is located on a prominent waterfront site and is the flagship of Hong Kong's trade on which the territory depends for its economic prosperity. High quality was provided.

(5) *Policy:* There was a transition from uncertainty on the part of the government into certainty on the part of the TDC once it had been given the responsibility to develop the project. The realisation stage eventually proceeded with a high level of certainty about what was required in principle. However there was uncertainty at the detailed level as the design ran parallel with construction which reduced the time available for decision-making. There was no apparent conflict on the project.

However, there were far more powerful environmental forces affecting the project which cannot even be allocated to these categories. They are more readily categorised at the macro-environmental level and are illustrative of Kast and Rosenzweig's (1985) view that the general environment can impinge directly on a project. Kast and Rosenzweig (1985) did not relate the phenomena to construction projects specifically in their account but the following does.

The instability of the project environment in the formative stages caused by weak economic conditions and lobbying by the TDC as a special interest group has been referred to earlier, particularly in relation to the changing client of the project. The account which follows considers the situation after that point.

When NWD entered into the contract for this project in 1984 great uncertainty prevailed regarding the reversion of Hong Kong to China, there was an oversupply of commercial property and the market was depressed. The Central District was the prime location on Hong Kong Island for commercial and hotel property. Nevertheless, NWD decided that, with the 'gift' of the site from the government, they could provide a world class convention and exhibition centre in exchange for the site and still make a profit. Their calculations must have been complex as assumptions had to be made about future income, as the two hotels and the office tower and service apartment block were to remain in their ownership. For NWD, this project was seen as a long-term investment.

By the time the project was completed in 1988, the property market had recovered and has remained buoyant ever since. It is considered that the site 'gifted' by the government in 1984 was worth up to five times this figure by 1988. NWD could provide the convention and exhibition area for approximately the value of the land with some margin. Even with these approximate figures it can be seen that in the run up to 1988 there was a great deal of money available in the project with which NWD could be generous to the TDC. This is particularly so as the value of the NWD real estate holdings on the site have escalated along with the increase in land value.

Primary and key decisions

The structure of this project is interesting in a number of respects. There was no conception system as, at the time of the decision to establish a CEC, it was also decided that real estate of some form was needed to house the CEC. Also a large number of key decisions were needed in the inception system due to strong environmental forces, as shown by the decision points in Table 13.7. Due to the complexity of the project the systems are not shown in as much detail as in the previous case studies.

Also of significance is the change in client during the project resulting in key decisions being taken by different organisations. The structures of client organisations have frequently been characterised as organisationally complex in terms of who wants the building, who will use it, who approves it, who provides the finance and other relevant responsibilities and powers (Cherns & Bryant 1984; Bennett 1985). This case study introduces a further significant complexity, that of changing clients each of which was organisationally complex. The government took key decisions nos 1–7, 9 and 10, the non-statutory board took key decisions no. 8 and the TDC key decisions nos 11–21. The lack of continuity between key decisions 6 and 7, 8 and 9 and between 9 and 10 will have reduced the effectiveness of the organisation structure of the project as knowledge and understanding were lost due to discontinuity. This situation is characterised by the establishment by the government of the non-statutory board for the purpose of identifying the location and form of the building, only to then reject the board's recommendation. The government then moved back into the role of client. It may be argued that the non-statutory board was part of the government system and not therefore a separate client but in fact its composition does not uphold this view.

Table 13.7 Decisions.

Primary decision
 TDC to be established
 A building may be required for an exhibition centre

Combined conception and inception system commences
 Key decision no. 1:   investigate suitability of the Hong Kong Stadium
 Key decision no. 2:   not to proceed with the Hong Kong Stadium
 Key decision no. 3:   investigate suitability of the World Trade Centre
 Key decision no. 4:   not to proceed with the World Trade Centre
 Key decision no. 5:   commission Earls Court study
 Key decision no. 6:   accept Earls Court study which recommended a new building

Realisation system commences
 Key decision no. 7:   appoint non-statutory board to process scheme for a new building
 Key decision no. 8:   a CEC to be developed on the Wanchai site recommended to the government
 Key decision no. 9:   the government reject recommendation of non-statutory board
 Key decision no. 10:   the government offer to allow the TDC to develop the project independently of the government and also offer the TDC the Wanchai site at no cost
 Key decision no. 11:   the TDC accept the government's offer
 Key decision no. 12:   C-Fin appointed project leaders
 Key decision no. 13:   indicative design and contract conditions agreed
 Key decision no. 14:   NWD appointed to design and construct the project on the basis of an interim agreement
 Key decision no. 15:   piling to commence ahead of formal agreement
 Key decision no. 16:   formal agreement signed which included the developed design
 Key decision no. 17:   Convention and Exhibition Centre, part of the development handed over to and accepted by the TDC
 Key decision no. 18:   completion of the remainder of the project (this was a phased process as the hotels, office tower and service apartments were commissioned; for simplicity they are not identified separately here)
 Key decision no. 19:   all contractual obligations discharged

---

Differentiation and integration of the operating system

Whilst the project is not as heavily differentiated as many, there still exists substantial differentiation particularly due to many of the consultants being located in different professional practices.

Hughes (1989) argues that differentiation on the basis of skill is the most significant factor and that skill diversity should match environmental complexity. This is a useful simplification but analysis on that basis would hide the particular features of this project for which reduced differentiation occurs because NWD, Polytown and Hip Hing are part of the same organisation and are therefore not differentiated by company and also because there are three architectural practices involved in the project which are therefore not differentiated by profession (or skill).

Differentiation is a natural phenomenon of managing complex projects and the level of differentiation should be matched by a corresponding provision of integration. The employment of a project management company (C-Fin) by the TDC and NWD's creation of its own in-house team (Polytown) as a separate entity was designed to achieve this and did so for nearly 100 per cent of the tasks they were involved with. However, the earlier tasks before C-Fin and Polytown were involved were not integrated and were characterised by lack of progress, discontinuity and a number of unsuccessful negotiations and reversal of decisions.

A key element in the success of the structure of this project was that NWD fully owned Polytown, their project managers and also Hip Hing, the main contractor. Hip Hing in turn owned the electrical and mechanical subcontractors. It was clear that within the NWD group all the companies served one master. For example, Hip Hing would do practically anything asked of them by NWD. There was no contract between Hip Hing and the parent company.

The key integrator was Polytown. Polytown as project manager for NWD, the developer, was strongly motivated to ensure that the project was successful. In this case the client (TDC) was not paying for the project directly. Payment was being made by allowing NWD to develop the hotels, office tower, etc., which were dependent for their success on the success of the Convention and Exhibition Centre. Hence, Polytown were committed not only to the success of NWD's parts of the project but equally to the success of the TDC's part.

What appears to be a complex project, which theoretically could be expected to create a highly differentiated organisation structure, actually had an organisation structure which reduced differentiation and provided a high level of integration. The facility to achieve such a structure is, to a large extent, provided by the contract strategy which allowed the TDC to subcontract the whole of the project to the developer, NWD.

Integration of the client

On this project the two primary integrators identified were:

(a) C-Fin, the project managers on behalf of TDC
(b) Polytown, the project managers on behalf of NWD.

The situation on this project is unusual because NWD is both client and developer. As a result Polytown are project managers on behalf of the client (NWD) for all parts of the development other than the actual CEC. In their other role Polytown are the project managers from the project team for the TDC's part of the project and C-Fin are acting in the conventional client representative project management role for the TDC's part of the project.

Unusual though the situation appears at first sight, it does in fact follow the theoretical position expected for client integration. As a result integration of the client was high in the realisation system. During the combined conception and inception system integration of the client was zero which had a significant effect on the pace at which the project progressed.

Differentiation of the managing and operating systems

Generally, the managing system and the operating system were separate on this project, although the project managers were directly involved in time planning. This is not unusual as none of the consultants undertake this activity as part of their normal services. A further aspect is that Polytown also undertook some architectural work but this was not done by people who were also project managing.

Complexity of the managing system within tasks was low in the realisation system. For many of the tasks of the inception system, a management system did not exist. As mentioned earlier serious discontinuity of the managing system at key decision points had an adverse effect on performance, particularly in the early stages.

Authority, power and leadership

Formal authority on the project vested in both the TDC as client for the CEC and NWD as client/developer of their parts of the project. Informal authority was significant on the project as the managing director of C-Fin had established a high reputation in project management achievements on the Mass Transit Railway and NWD had a high reputation as a developer as had many of their individual people in project management. Within this setting the project moved along smoothly in a climate of mutual respect and interests.

As a general backcloth it can be taken that the leadership style adopted by both project managers was a low-task high-relationship style (Blake & Mouton 1978). This observation accords with similar findings for Hong Kong (Rowlinson *et al.* 1993) in which it was found that project leaders in Hong Kong have more concern with maintaining good relationships and a harmonious working atmosphere than their Western counterparts.

At the operational level decisions were made in the normal manner by a member of Polytown, the project team and the contractor. Due to the integration of these contributors and the low-task high-relationship leadership style adopted there were no significant problems at this level. Whilst this leadership style was to a large extent a product of how the managers conducted themselves, certain mechanisms were built into the process to assist. It was written into the contracts of all the consultants that a senior member of each firm (in a named position) must attend a monthly meeting with C-Fin and Polytown to discuss progress and their firm's production. A similar group also met every month to look ahead at the project as a whole. This group broke down into smaller working parties which met more frequently.

The most distinguishing aspect was that every Wednesday night for two years the five main people from NWD and the lead architects met for dinner to discuss the project. A further feature was that every consultant had an employee permanently on site for the duration of the construction.

This disciplined structure and intense schedule of meetings ensured that the major players met regularly without fail, that feedback was swift and that concerted action was taken without delay.

The importance and high profile of the project meant that important decisions were to be made outside the formal authority structure. This proved to be the case when the decision was made to allow TDC to develop the project and such was the significance of the project, and the commitment of both TDC and NWD to its success, that when problems occurred during the design and construction of the project they were referred to the 'political' level for resolution. That this was quickly achieved by mutual adjustment shows that this constituted a legitimate use of power in the interests of achieving the objectives of the project.

It was at the key decision level, that is decisions taken by the client, that the process was most interesting. Whilst the TDC could be clearly seen as a client for the CEC, the position of NWD was as both developer and part-client. For most of the decisions which required consensus between these two, the employees of C-Fin, acting for the TDC, and of Polytown, acting for NWD, were able to resolve differences without too much difficulty. Where it was not possible to resolve differences at this level, they were discussed by the managing director of C-Fin and the managing director of NWD. Rather than keeping their distance from each other, they worked constructively towards a solution. The project was high profile and both individuals also had a high personal profile and were well respected members of the real estate industry in Hong Kong with a commitment to seeing the project succeed.

There were, however, a few instances when even they could not agree. At this stage, which on many projects would be leading to arbitration, the decision making moved up one step to the Chairperson of the TDC and, the owner-founder of NWD. Both were even more significant members of the Hong Kong community with an even greater personal interest in seeing the project succeed. The Chairperson of the TDC was the public figure most committed to the CEC and the co-founder of NWD saw the whole development as the flagship of NWD. This level of decision making was reached on very few occasions but on each occasion they made a decision which was implemented without dispute.

It is interesting to observe and recognise the leadership of this project above the project manager level (but nevertheless part of the project management process). Not only did this show the political nature of organisations as an internal phenomena, it also showed it to be an extremely important matter externally to organisations in preserving future business relationships. It forms the context within which project managers work but over which they have limited influence.

Conclusions

The process was divided into the conception and inception system together and the realisation system by a primary decision point and into the main sub-systems by key decision points. Whilst the organisation structure adopted at the realisation stage was novel, nevertheless, it closely subscribed to the propositions of the model to a much greater extent than conventional and other structures. The organisation structure will have contributed significantly to the successful outcome of the project particularly as it was complemented by effective authority patterns, appropriate leadership and positive power plays designed to achieve the objectives of the project.

The environmental influences on the project were particularly strong. Without a structure as effective as the one used it is doubtful whether they could have been overcome.

It could be argued that the government had been generous to the TDC and, hence, to NWD by allocating the land but if a broader view of the project is taken from the government perspective many other benefits accrued to the government. The Wanchai waterfront reclamation had just been completed, the property market was depressed, a significant and prestigious user was needed and the TDC needed a site. The cost to the government of reclamation was way below the value of the land. The multiplier effect of the decision to locate the TDC there is illustrated by the project which now surround it, including the 78 storey Central Plaza.

Once the project was handed to the TDC and, hence, to NWD it was risk free as far as the government was concerned and did not consume any government 'energy', which could be devoted to other affairs. The initial 'deal' was favourable as far as the government was concerned. It was only when the land value increased dramatically during construction that it looked as though the government had been generous. What really happened was that NWD were prepared to take the risk and were successful.

Environmental influences

Of significance is the manner in which macro-environmental forces impacted directly on the project. Whilst most of the literature sees such forces as remote from specific projects, insofar as they are filtered until micro-environmental forces actually reach the project system, political and economic forces were seen to directly affect the project system. The *political* aspects relate to the China/Britain agreement over Hong Kong and the consequent political decision by the Hong Kong Government to 'gift' the site in order to develop the Wanchai waterfront. The *economic* aspects are closely linked to the political as with increasing confidence after the signing of the Joint Declaration on Hong Kong by China and Britain the economy improved significantly making the project spectacularly successful against economic criteria.

The *cultural* aspects can be seen to be more than the micro-environmental issue of aesthetics. The Chinese approach to risk and their perception of social relationships influenced the readiness of the developer to undertake the project in initially adverse conditions and also determined the manner in which the project team were put together (Redding 1990). The latter overlaps with the *social* macro-environmental element.

This project exhibits the manner by which macro-environmental forces can impact much more directly on projects than is usually expected. Whilst one case study can do no more than raise questions it may well be that the degree to which macro-environmental forces impact on projects is a function of the size and significance of the project. If so, the question of how we measure size and significance remains.

As for macro-environmental forces, they were particularly strong and created high levels of uncertainty. In the early stages of the project they were negative as political and economic conditions were difficult. The organisation was not structured to cope with such forces as it lacked a management system and the necessary differentiation and corresponding integration. It did not have the leadership necessary to put such mechanisms in place and was characterised by changing clients which created discontinuity at key decisions. The result was that 18 years elapsed before the project realisation system commenced.

The contrast with the management of the realisation stage was dramatic. Once the TDC had been given responsibility for developing the project and appointed their project managers, an organisational structure was put in place which was capable of dealing with the still powerful negative macro-environmental forces.

Attributes

There was a high level of satisfaction with the completed project which had a high correlation of the project organisation's configuration with the model's propositions. Any implied criticism of the performance of the process should be mitigated by the high measure of success achieved in the project outcome in difficult circumstances.

Chapter 14
Postscript

In looking to future developments in project management theory, the main area for attention may be organisation culture. Organisation culture is well recognised in the general management literature but how it may manifest in a project setting, if at all, has not been explored in any depth.

Organisation culture is understood to be the shared beliefs and values that members of an organisation hold in common, resulting in a predictable pattern of general behaviour. It sets the tone of the organisation and establishes implied rules for the way in which people behave (Weihrich & Koontz 1994). Such a culture develops incrementally as people accept the prevailing values and, once established, resists change. Scott (1992) considers that organisations with a strong culture rely primarily on an informal structure and on individuals embracing common norms and values that orient and govern their contribution and that such a culture acts as an internal control on the behaviour of the members of the organisation.

Scott recognises that every organisation has a culture but that cultures vary in their attributes. Focus has been on organisations developing 'strong' cultures as advantageous to business organisations. But such 'strong' cultures, which rely on unwavering belief in the company, tend towards an authoritarian style which may gain benefit in the short term through enhanced commitment but may in the long term be inhibited by resistance to change. Such cultures sit uncomfortably in societies based on pluralism and committed to democratic institutions. He points out that what should be emphasised is the diversity and variety of organisation culture using Meyerson and Martin's (1987) work which:

> '... contrasts dominant approaches which stress the unity of cultural beliefs within an organisation with others that stress the extent of differentiation – the identification of subcultures – or still others that acknowledge the absence of a shared, integrated set of values: a culture recognising ambiguity. Meyerson and Martin treat these differences as paradigms – analytic models applicable to any organisation – but also recognise that the cultures of specific organisations may be better characterised by one rather than another perspective.'

The question is what are the implications, if any, of such considerations for construction project organisations? The individual firms which make up the temporary project organisations will each have their own corporate culture. This can be seen to be represented in part by the 'technology' element of differentiation incorporated in the model developed in this book together with the sentience which exists within construction project firms. To this will be added the distinctive culture which may have developed within each firm. At present there appears to be no formal understanding of the range of cultures which exist in construction related companies, nevertheless the challenges to the project manager in integrating such a potentially diverse group of firms is clear to see.

The question is what is the nature of the culture of the project organisation? This question assumes that it is possible for a temporary project organisation to have an organisation culture. Based on the definition of organisation culture which stresses commonality, the issue is how much commonality is developed with such transient organisations. Following from this is the question of whether commonality is beneficial to project teams and if so in what degree and to what extent. A first reaction may be that the commitment which a 'strong' culture brings would be beneficial and may be more likely to occur on long duration projects and/or those with little differentiation amongst project team members but on the other hand such a 'strong' culture may inhibit creativity and flexibility in devising solutions. Conversely, short duration projects with a highly differentiated team may have a culture which does not value commitment and thereby have a relatively unmotivated project team. To what extent should a project manager aim to build an organisation culture and of what variety?

These questions are as yet unaddressed by research but are shown to be recognised as important by industry as evidenced by the growth in partnering and other relational contracting ideas as evaluated by Alsagoff and McDermott (1994). They undertook a survey to indicate the extent of relational contracting in the UK and whilst over 35 per cent of clients claimed long-term relationships with contractors, the researchers concluded that relational contracting had only been taken up partly and is continually limited by the volatile nature of the industry and the adversarial attitudes of its constituents. Clearly, although partnering and other relational contracting approaches had been adopted, a project culture which generated the benefit expected by such approaches had not been achieved in the majority of cases.

The focus so far has been on organisation culture within firms and project organisations but the wider impact of culture is indicated by Scott when he points out that anthropologists have long believed most of the orderliness and patterning found in social life is accounted for by cultural systems, but that such views have only recently been employed by students of organisations. Whilst local projects undertaken by local consultants and contractors have to contend with different organisation cultures, international projects undertaken by consultants and contractors from many different countries, often for clients from different countries which could be located in yet another different country, will generate organisation cultures which may vary by a further order of magnitude from local projects. The management of such projects poses the need for a high level of understanding of the diverse traditional cultures embedded in the contributors to the project as a result of their country of origin, and, if appropriate, a huge challenge in creating a project organisation culture.

Understandably, empirical work on international cross cultural construction projects is limited due to the sheer scale of the research needed to obtain useful results. This is due not only to the magnitude of the task of obtaining data over what would probably need to a long period but more importantly because there is as yet no theoretical basis to the analysis of organisation culture. What does exist is a range of publications which have identified some of the issues which require the attention of managers of multi-national enterprises only some of which relate directly to construction. Not all are reviewed here, only sufficient to give an indication of the nature of the issues to be confronted.

Winch and Campagnae's (1995) work, although not directed at culture specifically, shows that even in countries which are neighbours distinct differences occur in the basis on which construction projects are organised. In particular the professional orientation of the British system generates a culture which is distinctly different from France's greater reliance on the contractor. A further contrast with a direct implication for culture is the way in which the French contractor internalises variability by retaining a large direct labour force rather than externalising variability by subcontracting as is the case in Britain.

Wang (1992) found that some Chinese cultural traditions:

'had a strong influence on work behaviour in joint ventures. Among the traditions, equality and egalitarianism were popular among many employees who, for example, were not not comfortable about distinctive wage differences at different organisational levels. In particular, some joint ventures had to follow the conventional Chinese working style and management behaviour of non-joint-venture companies. a phenomenon called "Chinese nationalisation of joint-venture management". Also, the traditional value of emphasising interpersonal relationships and group responsibility was still quite dominating among Chinese employees.'

He also found that decision-making styles in joint ventures (not construction) were largely influenced by the management traditions of the managers' own countries. Redding (1990) also identified the fundamental beliefs and values of Chinese tradition in his study of the overseas Chinese in business. Much is seen to be drawn from Confucianism including familism, obedience, piety and collectivism.

Whilst such cultural forces are strong, Clegg (1990) offered both 'cultural' and 'economic' explanations for the growth of the economies in South East Asia and concluded that 'the success of Asian business cannot be attributed so simply to cultural factors despite the stress on these in the literature'. He considered culturalist explanations in Confucian terms too general and unspecific and thought that whilst culture can explain some common patterns across the East Asian societies, it cannot explain variations within and between them.

The difficulty of dealing with these kinds issues in a rigorous research sense is further illustrated by Child and Markoczy's (1993) study of Chinese and Hungarian joint ventures. They found that hierarchy and reciprocal relationships were major elements shared by Chinese and Hungarian cultures and are consistent with the relatively passive behaviour of local managers and the fact that they approached relations in such a way as to sustain dependencies. They found 'face' and collectivism stronger amongst the Chinese. Nevertheless they concluded that 'while national cultures almost certainly generate predispositions towards certain behavioural patterns, it is evident that system differences within the same cultural domain have an overriding impact on managerial practices'.

The reason for quoting these accounts of leading general management scholars on organisation culture issues is to emphasise how confusing the picture appears and to highlight the lack of an analytical framework for understanding the effect of culture on organisation performance. In the specific field of project management even less is documented about the effect of culture, either organisation culture or traditional cultural values. Yet construction projects are most vulnerable to dysfunction due to the effects of both types of cultural phenomena. On the one hand by being subject to the organisation culture of firms together with the difficulty of developing a project organisation culture and on the other by needing to cope with the traditional culture of the contributing groups within the cultural context of the country which is host to the project. The dearth of research is illustrated by one study in the area (Gray *et al.* 1990) which did not address culture directly but focused on organisation structure. It covered construction and other project organisations and found that project managers working on similar kinds of projects attach similar priorities to project objectives; but all the project managers were from developed countries.

The purpose of this postscript is to attempt to indicate what may possible be the next main thrust of development of project management theory. The framework presented in this book provides a basic analytical structure for understanding the effectiveness of project organisations but much remains to be explained. Much of that explanation may be in a better understanding of organisation culture in all its facets but it is a complex yet elusive subject. It may be that organisation culture is in reality the totality of organisation behaviour and therefore not separately identifiable, hence its appearance of elusivity. It may be everything or it may be nothing. Maybe the behaviour of organisations in all their aspects (if ever we are able to understand them all) needs to be identified within a framework which complements the organisational propositions in this book, which, when summed, represents culture (both organisational and traditional). One thing is certain, there is much work to do.

That the traditional culture of the participants to construction projects needs to be taken into account in the manner in which they are managed is clear to many project managers, for example the traditionally adversarial approach of Western contractors as against the more conciliatory approach of their Eastern counterparts. Yet the impact of such differences is not understood. The effect of organisation culture on project organisation is even less understood and may be found to be an inappropriate paradigm for analysis of project organisations but we do not yet know.

Interestingly, the Latham Report (Latham 1994) was concerned almost totally with technical and procedural matters such as client briefing, project information and contractual matters. Important and valuable as these are in terms of making the system more effective, they do not address the underlying characteristics of the system. There is no reference in the report's recommendations to changing the attitudes of the participants towards less adversarial postures. In fact the only suggestion on this count seems to be that 'endlessly refining existing conditions of contract will not solve adversarial problems'. Maybe at a practical level, organisation culture is not seen as a significant issue yet the interest in partnering and other relational approaches and their adoption seem to belie this view.

Productive initiatives in furthering our understanding of the project management process is likely to come from the application of the behavioural and psychological disciplines to the project management process. Only then is further understanding of issues, such as co-operation, collaboration, power and motivation, likely to emerge. Examples of new directions in the application of these disciplines to construction can be found in Liu's (1994) modelling of construction procurement as an act-to-outcome process model of organisational behaviour, Raftery's (1994) attention to the human aspects of project risk management after years of focus on techniques involved and Rowlinson *et al.*'s (1993) examination of leadership styles in mixed culture construction projects which showed that project leaders in Hong Kong have more concern with maintaining good relationships and a harmonious working atmosphere than those in the West.

We may need to pull back from the all-embracing idea of culture to examine the specific elements of behaviour which may make up such a culture, whether from an organisational or traditional culture base. We will see what the next few years bring.

References

Ackoff, R.L. (1969) Systems, organisation and interdisciplinary research. In: Emery, F.E. (ed) *Systems Thinking*. London: Penguin.

Ackoff, R.L. (1971) Towards a system of system concepts, *Management Science*, **17**, 661–71.

Agger, B. (1991) Critical theory, poststructuralism, postmodernism: their sociological relevance. *Annual Review of Sociology*, **17**, 105–31.

Akintoye, A. (1994) Design and build: a survey of construction contractors' views. *Construction Management and Economics*, **12**, 155–63.

Allen, D. (1984) Towards the client's objective. In: Brandon, P.S. & Powell, J.A. (eds) *Quality and Profit in Building Design*. London: Spons.

Alsagoff, A. & McDermott, P. (1994) Relational contracting: a prognosis of the UK construction industry. In: Rowlinson, S.M. (ed) *Proceedings of CIB W92 Symposium: East Meets West: Procurement Systems*. CIB Publication No. 175. Hong Kong: Department of Surveying, University of Hong Kong.

Anderson, S.D. (1992) Project quality and project managers. *International Journal of Project Management*, **10**, 138–44.

Anon (1955) How to know who does what. *Mill and Factory*, January, 75–8.

Association of Project Managers (1984) *Closing the Gaps in Project Management Systems*. London: Butterworths.

Barnard, I. (1938) *The Functions of the Executive*. Cambridge, MA: Harvard University Press.

Barney, J.B. (1990) The debate between traditional management theory and organisational economics: substantive differences or intergroup conflicts? *Academy of Management Review*, **15**, 382–93.

Bennett, J. (1985) *Construction Project Management*. Cambridge: Butterworth.

Bennett, J. (1991) *International Construction Project Management: General Theory and Practice*. Oxford: Butterworth-Heinemann.

Bennett, J. (1994) Book review – The Management of Projects. *Construction Management and Economics*, **12**, 279–80.

Bennigson, L.A. & Balthasas, H.V. (1974) Forecasting coordination problems in pharmaceutical research and development. In: *1974 Proceedings of the Project Management Institute*. Paris: Internet.

Bennis, W.G. (1959) Leadership theory and administrative behaviour. *Administrative Science Quarterly*, **4**, 2590–301.

Bertalanffy, L. von (1969) *General Systems Theory: Essays on its Foundation and Development*. New York: Braziller.

Bishop, D. (1968) *The Background to Management Studies by the BRS*. Building Research Station Current Paper 60/68. London: HMSO.

Blake, R.R. & Mouton, J.S. (1978) *The New Managerial Grid*. Houston: Gulf Publishing.

Bonoma, T.V. & Slevin, D. (1978) *Executive Survival Manual*. Belmont: Wadsworth Publishing Co.

Boulding, K.E. (1956) General systems theory: the skeleton of science. *Management Science*, **2** 197–208.

Bowley, M. (1966) *The British Building Industry*. Cambridge: Cambridge University Press.

Bresnen, M.J. (1991) Construction contracting in theory and practice: a case study. *Construction Management and Economics*, **9**, 247–63.

Bresnen, M.J. & Haslam, C.D. (1991) Construction industry clients: a survey of their attributes and project management practices. *Construction Management and Economics*, **9**, 327–42.

British Property Federation (1983) *Manual of the BPF System*. London: British Property Federation.

Buchko, A. (1994) Conceptualisation and measurement of environmental uncertainty: an assessment of the Miles and Snow Perceived Environmental Uncertainty Scale. *Academy of Management Journal*, **37**, 410–25.

Buckley, W. (1968) 'Society as a complex adaptive system: In: Buckley, W. (ed.) *Modern Systems Research for the Behavioural Scientist: A Sourcebook*. Chicago: Aldine.

Burke, W.W. (1986) Leadership as empowering others. In: Srivastna, S. (ed.) *Executive Power*. San Francisco: Jossey-Bass.

Burns, R. & Stalker, G.M. (1966) *The Management of Innovation*. London: Tavistock Publications.

Campbell, J.P., Dunnette, M.D., Lawler, E.E., Weick, K.E., (1970) *Managerial Behaviour, Performance and Effectiveness*. New York: McGraw-Hill.

Carpenter, J.B.G. (1981) The UK system of construction procurement and what is wrong: how to improve? Paper to *RICS Quantity Surveyor's 12th Triennial Conference*. London: RICS.

Chartered Institute of Building (1979) *Project Management in Building*. Occasional Paper No. 20. Ascot: The Chartered Institute of Building.

Chau, K.W. & Walker, A. (1994) Institutional costs and the nature of subcontracting in the construction industry. In: Rowlinson, S.M. (ed.) *Proceedings of CIB W92 Symposium: East Meets West: Procurement Systems*. CIB Publication No. 175. Hong Kong: Department of Surveying, University of Hong Kong.

Cherns, A.B. & Bryant, D.T. (1984) Studying the client's role in construction management. *Construction Management and Economics*, **2**, 177–84.

Child, J. (1972) Organisational structure, environment and performance. The role of strategic choice. *Sociology*, **6**, 1–22.

Child, J. (1977) *Organisation*. London: Harper & Row.

Child, J. & Markoczy, L. (1993) Host country managerial behaviour and learning in Chinese and Hungarian joint ventures. *Journal of Management Studies*, **30**, 611–32.

Clegg, S.R. (1990) *Modern Organisation*. London: Sage.

Cleland, D.I. & King, W.R. (1972) *Management: A Systems Approach*. New York: McGraw-Hill.

Cleland, D.I. & King, W.R. (1983) *Systems Analysis and Project Management*. New York: McGraw-Hill.

Co-ordinated Project Information (1987) *Co-ordinated Project Information for Building Works, a Guide with Examples*. London: Construction Project Information (CPI).

Coase, R.H. (1937) The nature of the firm. *Economica*, n.s., **4**, 386–405.

Coase, R.H. (1988) *The Firm, the Market and the Law*. Chicago: The University of Chicago Press.

Conger, J.A. & Kanungo, R.N. (1988) The empowerment process: integrating theory and practice. *Academy of Management Review*, **13**, 471–82.

Construction Industry Institute (1989) *Partnering: Meeting the Challenges of the Future*. Texas: Construction Industry Institute (CII).

Dahlman, C.J. (1979) The problem of externality. *The Journal of Law and Economics*, **22**, 141–62.

Dalton, G.W., Lawrence, P.C. & Lorsch, J.W. (1970) *Organisational Structure and Design*. New York: Irwin.

Daniel, D.W. (1990) Hard problems in a soft world. *International Journal of Project Management*, **8**, 79–83.

Department of Industry (1982) *The United Kingdom Construction Industry*. London: HMSO.

Department of the Environment (1995) *Housing and Construction Statistics*. London: HMSO.

DiMaggio, P. & Powell, W. (1983) The iron cage revisited: institutional isomorphism and collective rationality in organisation fields. *American Sociological Review*, **48**, 147–60.

Donaldson, L. (1990a) The etheral hand: organisational economics and management theory. *Academy of Management Review*, **15** 369–81.

Donaldson, L. (1990b) A rational thesis for criticisms of organisational economics: a reply to Barney. *Academy of Management Review*, **15**, 394–401.

Dornbusch, S.M. & Scott, W.R. (1975) *Evaluation and the Exercise of Authority*. San Francisco: Jossey-Bass.

Eilon, S. (1979) *Aspects of Management*. Oxford: Pergamon Press.

Emerson, R.M. (1962) Power-dependence relations. *American Sociological Review*, **27**, 31–40.

Emery, F.E. (1959) *Characteristics of Socio-Technological Systems*. Tavistock Document 527. London: Tavistock Publications.

Emery, F.E. & Trist, E.L. (1965) The causal texture of organisational environment. *Human Relations*, **18**, 21–32.

Fayol, H. (1949 trans.) *General and Industrial Management*. London: Pitman (First published in 1919).

Fielder, F.E. (1967) *A Theory of Leadership Effectiveness*. New York: McGraw-Hill.

Finkelstein, S. (1992) Power in top management teams: dimensions, measurement and validation. *Academy of Management Journal*, **35**, 505–38.

French, W. & Bell, C. (1990) *Organisation Development*. Englewood Cliffs, NJ: Prentice-Hall.

Fryer, B. (1985) *The Practice of Construction Management*. Oxford: Blackwell Science.

Gilbert, G.P. (1983) Styles of project management. *International Journal of Project Management*. **1**, 189–93.

Gray, C., Dworatschek, S., Gobeli, D., Kroepfel, H. & Larsa, E. (1990) International comparison of project organisation structures: use and effectiveness. *International Journal of Project Management*, **8**, 26–32.

Green, S.D. (1994) Sociological paradigms and building procurement. In: Rowlinson, S.M. (ed.) *Proceedings of CIB W92 Symposium: East Meets West: Procurement Systems*. CIB Publications No. 175. Hong Kong: Department of Surveying, University of Hong Kong.

Greening, D.W. & Gray, B.L. (1994) Testing a model of organisational response to social and political issues. *Academy of Management Journal*, **37**, 467–98.

Griesinger, D.W. (1990) The human side of economic organisation. *Academy of Management Review*, **15**, 478–99.

Gulick, L. & Urwick, L. (eds) (1937) *Papers on the Science of Administration*. New York: Institute of Public Administration, Columbia University.

Gunnarson, S. & Levitt, R.E. (1982) Is a building construction project a hierarchy or a market? In: Riis, J.O. (ed.) *Proc. 7th World Congress on Project Management*. Copenhagen: Internet.

Handler, A.B. (1970) *Systems Approach to Architecture*. Amsterdam: Elsevier.

Hennart, J.F. (1994) The comparative institutional theory of the firm: some implications for corporate strategy. *Journal of Management Studies*, **31**, 193–207.

Hersey, P. & Blanchard, K.H. (1972) *Management of Organisation Behaviour*. New Jersey: Prentice-Hall.

Herzberg, F. (1968) *Work and the Nature of Man*. London: Staples Press.

Hesterly, W.S., Liebestkind, J. & Zenger, T.R. (1990) Organisational economics: an impending revolution in organisation theory? *Academy of Management Review*, **15**, 402–420.

Hickson, D.J., Hinings, C.R., Lee, C.A., Schreck, R.E. & Pennings, J.M. (1971) A strategic contingencies' theory of intraorganisational power. *Administrative Science Quarterly*, **16**, 216–29.

Higgins, G. & Jessop, N. (1965) *Communication in the Building Industry*. London: Tavistock Publications.

HMSO (1944) *Report of the Management and Planning of Contracts (The Simon Report)*. London: HMSO.

HMSO (1950) *Report of the Working Party on the Building Industry (The Phillips Report)*. London: HMSO.

HMSO (1962) *Survey of the Problems Before the Construction Industries (The Emmerson Report)*. London: HMSO.

HMSO (1964) *The Placing and Management of Contracts for Building and Civil Engineering Works (The Banwell Report)*. London: HMSO.

HMSO (1967) *A survey of the Implementation of the Recommendations of the Committee under the Chairmanship of Sir Harold Banwell on the Placing and Management of Contracts for Building and Civil Engineering Works (Action on the Banwell Report)*. London: HMSO.

Hughes, W.P. (1989) *Organisational Analysis of Building Projects*. PhD thesis, Liverpool: Liverpool Polytechnic (now Liverpool John Moores University).

Ireland, V. (1985) The role of managerial actions in the cost, time and quality performance of high rise commercial building projects. *Construction Management and Economics*, **3**, 59–87.

Janis, I.L. (1983) *Groupthink*. Boston: Houghton Mifflin.

Jensen, M.C. & Meckling, W.Y. (1976) Theory of the firm: managerial behaviour, agency costs, and ownership structure. *Journal of Financial Economics*, **3**, 305–60.

Kanter, R.M. (1977) *Men and Women of the Corporation*. New York: Basic Books.

Kanter, R.M. (1983) *The Change Masters*. London: Unwin.

Kast, F.E. & Rosenzweig, J.E. (1985) *Organisation and Management: A Systems and Contingency Approach*. New York: McGraw-Hill.

Katz, D. & Kahn, R.L. (1978) *The Social Psychology of Organisations*. New York: John Wiley & Sons.

Knowles, R. (1986) Project manager: the legal position. *The Chartered Quantity Surveyor*. July (Part 1), 11; August (Part 2), 10–11.

Knox, F. & Hennesey, J. (1966) *Restrictive Practices in the Building Industry*. London: Institute of Economic Affairs.

Kometa, S.T., Olomolariye, P.O. & Harris, F.C. (1994) Attributes of UK construction clients influencing project consultants performance. *Construction Management and Economics*, **12**, 433–43.

Koontz, H. (1961) The management theory jungle. *Journal of the Academy of Management*, December.

Lansley, P. (1994) Analysing construction organisations. *Construction Management and Economics*, **12**, 337–48.

Larson, E. (1995) Project partnering: results of study of 280 construction projects. *Journal of Management in Engineering*, **11**, 30–35.

Latham, M. (1994) *Constructing the Team*. London: HMSO.

Lawrence, P.C. & Lorsch, J.W. (1967) *Organisation and Environment: Managing Differentiation and Integration*. Boston: Graduate School of Business Administration, Harvard University.

Leigh, A. (1983) *Decisions, Decisions! A Practical Guide to Problem Solving and Decision Making*. Aldershot: Gover.

Likert, R. (1961) *New Patterns of Management*. New York: McGraw-Hill.

Liu, M.M.A. (1994) From act to outcome – a cognitive model of construction procurement. In: Rowlinson, S.M. (ed.) *Proceedings of CIB W92 Symposium: East Meets West: Procurement Systems*. CIB Publication No. 175. Hong Kong: Department of Surveying, University of Hong Kong.

Liu, M.M.A. (1995) *Evaluation of the Outcome of Construction Projects*. PhD Thesis, Hong Kong: University of Hong Kong.

Loosemore, M. (1994) Problem behaviour. *Construction Management and Economics*, **12**, 511–20.

Lovell, R.J. (1993) Power and the project manager. *International Journal of Project Management*, **11**, 73–8.

Maloney, W.F. & Federle, M.O. (1991) Organisation, culture and management. *Journal of Management in Engineering*, **7**, 43–58.

Manz, C.C. & Neck, C.P. (1995) Teamthink: beyond the groupthink syndrome in self-managing work teams. *Journal of Managerial Psychology*, **10**, 7–15.

March, T.G. & Simon, H.A. (1958) *Organisation*. New York: John Wiley & Sons.

Maslow, A.H. (1954) *Motivation and Personality*. New York: Harper.

Masterman, J.W.E. & Gameson, R.N. (1994) Client characteristics and needs in relation to their selection of building procurement systems. In Rowlinson, S.M. (ed.) *Proceedings of CIB 92 Symposium: East Meets West: Procurement Systems*. CIB Publication No. 175. Hong Kong: Department of Surveying, University of Hong Kong.

McGregor, D. (1960) *The Human Side of Enterprise*. New York: McGraw-Hill.

Meyer, J.W. & Rowan, B. (1977) Institutionalised organisations: formal structure as myth and ceremony. *American Journal of Sociology*, **83**, 340–63.

Meyerson, D. & Martin, J. (1987) Cultural change: an integration of three different views. *Journal of Management Studies*, **24**, 623–47.

Miles, R.E. & Snow, C.C. (1978) *Organisational Strategy, Structure, and Process*. New York: McGraw-Hill.

Miller, E.J. (1959) Technology, territory and time: the internal differentiation of complex production systems. *Human Relations*, **12**, 243–72.

Miller, E.J. & Rice, A.K. (1967) *Systems of Organisation: the Control of Task and Sentient Boundaries*. London: Tavistock Publications.

Mintzberg, H. (1973) *The Nature of Managerial Work*. New York: Harper and Row.

Mintzberg, H. (1979) *The Structure of Organisations*. Englewood Cliffs, N.J.: Prentice-Hall.

Mintzberg, H. (1989) *Mintzberg on Management: Inside Our Strange World of Organisations*. New York: Free Press.

Mobbs, G.N. (1976) *Industrial Investment – A Case Study in Factory Building*. Slough: Slough Estates Ltd.

Mohsini, R.A. & Davidson, C.H. (1992) Determinants of performance in the traditional building process. *Construction Management and Economics*, **10**, 343–59.

Morris, P.W.G. (1972) *A Study of Selected Building Projects in the Context of Theories Organisation*. PhD Thesis, Manchester: University of Manchester, Institute of Science and Technology.

Morris, P.W.G. (1994) *The Management of Projects*. London: Thomas Telford.

Nahapiet, H. & Nahapiet, J. (1985) A comparison of contractual arrangements for building projects. *Construction Management and Economics*, **3**, 217–31.

Napier, J.A. (1970) *A Systems Approach to the Swedish Building Industry*. Stockholm: The National Swedish Institute for Building.

National Economic Development Council (1964) *The Construction Industry*. London: HMSO.

National Economic Development Office (1975) *The Public Client and the Construction Industries*. London: HMSO.

National Economic Development Office (1976) *The Professions in the Construction Industries.* London: HMSO.

National Economic Development Office (1978) *Construction for Industrial Recovery.* London: HMSO.

National Economic Development Office (1983) *Faster Building for Industry.* London: HMSO.

National Economic Development Office (1985) *Thinking about Building.* London: HMSO.

National Economic Development Office (1987) *Faster Building for Commerce.* London: HMSO.

Naylor, J., Pritchard, R.D. & Ilgen, D.R. (1980) *A Theory of Behaviour in Organisations.* New York: Academic Press.

Neilsen, E. (1986) Empowerment strategies: balancing authority and responsibility. In: Sirwastra, S. (ed.) *Executive Power.* San Francisco: Jossey-Bass.

Newcombe, R. (1994) Procurement paths – a power paradigm. In: Rowlinson, S.M. (ed.) *Proceedings of CIB W92 Symposium: East Meets West: Procurement Systems.* CIB Publication No. 175. Hong Kong: Department of Surveying, University of Hong Kong.

Peter, L.J. & Hull, R. (1969) *The Peter Principle.* New York: Morrow.

Pfeffer, J. (1978) The micro politics of organisations. In: Meyer, M.W. (ed.) *Environments and Organisations.* San Francisco: Jossey-Bass.

Pfeffer, J. (1981) *Power in Organisations.* Marshfield, MA: Pitman.

Pfeffer, J. (1992) *Managing with Power: Politics and Influence in Organisations.* Boston, MA: Harvard Business School Press.

Pfeffer, J. & Salanick, G.R. (1978) *The External Control of Organisations.* New York: Harper & Row.

Poirot, J.W. (1991) Organising for quality: matrix organisation. *Journal of Management in Engineering,* **7**, 178–86.

Porter, L.W. & Lawler, E.E. (1968) *Managerial Attitudes and Performance.* Homewood, IL: Dorsey.

Quinn, R.E. (1988) *Beyond Rational Management: Mastering the Paradoxes and Competing Demands of High Performance.* San Francisco: Jossey-Bass.

Raftery, J. (1994) Human aspects of project risk management. In: Rowlinson, S.M. (ed.) *Proceedings of CIB W92 Symposium: East Meets West: Procurement Systems.* CIB Publication No. 175. Hong Kong: Department of Surveying, University of Hong Kong.

Redding, S.G. (1990) *The Spirit of Chinese Capitalism.* New York: Walter de Gruyter.

Reve, T. & Levitt, R.E. (1984) Organisation and governance in construction. *International Journal of Project Management,* **2**, 17–25.

Ries, C.J. (1964) *The Management of Defense.* Baltimore: John Hopkins Press.

Roberts, C.C. (1989) Intermittent project management. *Journal of Management in Engineering,* **5**, 84–9.

Robins, M.J. (1993) Effective project management in a matrix-management environment. *International Journal of Project Management,* **11**, 11–14.

Rouleau, L. & Sequin, F. (1995) Strategy and organisation theories: common forms of discourse. *Journal of Management Studies,* **32**, 22–38.

Rowlinson, S.M. (1988) *An Analysis of Factors Affecting Project Performance in Industrial Building.* PhD Thesis, Uxbridge: Brunel University.

Rowlinson, S., Ho, T.K.K. & Yuen, P.H. (1993) Leadership style of construction managers in Hong Kong. *Construction Management and Economics,* **11**, 455–65.

Royal Institute of British Architects (1991) *Handbook of Architectural Practice and Management.* London: Royal Institute of British Architects.

Scott, W.R. (1992) *Organisations: Rational, Natural and Open Systems.* Englewood Cliffs, NJ: Prentice-Hall.

Seifers, L. von (1972) *A Contingency Theory Approach to Temporary Management Sys-*

tems. PhD Thesis, Cambridge, Mass: Graduate School of Business Administration, Harvard University.

Slevin, D.P. (1983) Leadership and the project manager. In: Cleland, D.I. & King, W.R. (eds) *Project Management Handbook*. New York: Van Nostrand Reinhold.

Simon, H.A. (1947) *Administrative Behaviour*. New York: Free Press.

Smith, K.G., Carroll, S.J. & Ashford, S.J. (1995) Intra and interorganisational cooperation: towards a research agenda. *Academy of Management Journal*, **38**, 7–23.

Tannenbaum, R. & Schmidt, W.H. (1973) How to choose a leadership pattern. *Harvard Business Review*, **51**, 162–80.

Tavistock Institute (1966) *Interdependence and Uncertainty*. London: Tavistock Publications.

Taylor, F.W. (1911) *The Principles of Scientific Management*. New York: Harper.

Thompson, J. (1967) *Organisation in Action*. New York: McGraw-Hill.

Thompson, N. (1978) Alternative methods of management. *Building*, 27 January, 67–70.

Thompson, P. (1991) The client role in project management. *International Journal of Project Management*, **9**, 90–92.

Tolbert, P.S. & Zucker, L.G. (1983) Institutional sources of change in the formal structure of organisations: the diffusion of civil service reforms, 1880–1935. *Administrative Science Quarterly*, **28**, 22–39.

Townsend, R. (1984) *Further Up The Organisation*. London: Michael Joseph.

Turner, J.R. (1994) Editorial: Project management: Future developments for the short and medium terms. *International Journal of Project Management*, **12**, 3–4.

Vroom, V.H. (1964) *Work and Motivation*. Malabar, FL: Robert E. Krieger Publishing Co.

Walker, A. (1976) *Project Management: A Review of the State of the Art*. London: The Institute of Quantity Surveyors (now the Royal Institution of Chartered Surveyors).

Walker, A. (1980) *A Model of the Design of Project Management Structures for Building Clients*. PhD Thesis, Liverpool: Liverpool Polytechnic (now Liverpool John Moores University).

Walker, A. (1994) *Building the Future – The Story of the Controversial Construction of the Campus of the Hong Kong University of Science and Technology*. Hong Kong: Longman Asia Ltd.

Walker, A. (1995) *Hong Kong: The Contractors' Experience*. Hong Kong: Hong Kong University Press.

Walker, A. & Hughes, W.P. (1984) Private industrial project management: A systems-based case study. *Construction Management and Economics*, **2**, 93–110.

Walker, A. & Hughes, W.P. (1986) A conventionally-managed project: a systems-based case study. *Construction Management and Economics*, **4**, 57–74.

Walker, A. & Hughes, W.P. (1987a) A project managed by a multidisciplinary practice: a systems-based case study. *Construction Management and Economics*, **5**, 123–40.

Walker, A. & Hughes, W.P. (1987b) An analysis of the management of a public sector project: a systems-based case study. In: Lansley, P. & Harlow, P. (eds) *Managing Construction Worldwide*. London: Spons.

Walker, A. & Kalinowski, M. (1994) An anatomy of a Hong Kong project – organisation, environment and leadership. *Construction Management and Economics*, **12**, 191–202.

Walker, A. & Wilson, A.J. (1983) An approach to the measurement of the performance of the project management process. In: Chiddick, D. & Millington, A. (eds) *Proceedings of the Land Management Research Conference*. London: Spon.

Walker, D.H.T. (1994) Procurement systems and construction time performance. In: Rowlinson, S.M. (ed.) *Proceedings of CIB W92 Symposium: East Meets West: Procurement Systems*. CIB Publication No. 175. Hong Kong: Department of Surveying, University of Hong Kong.

Wang, Z.M. (1992) Managerial psychological strategies for Sino-foreign joint ventures. *Journal of Managerial Psychology*, **7**, 10–16.

Ward, S.C., Curtis, B. & Chapman, C.B. (1991) Objectives and performance in construction projects. *Construction Management and Economics*, **9** 343–53.

Weber, M. (1947 trans.) *The Theory of Social and Economic Organisation.* (ed. A.H. Henderson & T. Parsons). Glencoe, IL: Free Press (first published in 1924).

Weber, M. (1968 trans.) *Economy and Society: An Interpretive Sociology*, 3 vols (eds. Guenther Roth & Claus Wittich). New York: Bedminster Press, (first published in 1924).

Weihrich, H. & Koontz, H. (1993) *Management: A Global Perspective.* New York: McGraw-Hill.

Williamson, O.E. (1975) *Markets and Hierarchies: Analysis and Antitrust Implications.* New York: Free Press.

Williamson, O.E. (1979) Transaction-cost economics: the governance of contractual relations. *The Journal of Law and Economics*, **22**, 233–61.

Williamson, O.E. (1981a) The economics of organisation and the transaction cost approach. *American Journal of Sociology*, **87**, 548–77.

Williamson, O.E. (1981b) The modern corporation: origins, attributes. *Journal of Economic Literature*, **19**, 1537–68.

Williamson, O.E. (1985) *The Economic Institutions of Capitalism.* New York: Free Press.

Wilson, A.J. (1976) Thirteenth century project management. *Building Technology and Management*, **7**, 5–8.

Winch, G. (1989) The construction firm and the construction project: a transaction cost approach. *Construction Management and Economics*, **7**, 331–45.

Winch, G. & Campagnae, E. (1995) The organisation of building projects: an Anglo/French comparison. *Construction Management and Economics*, **13**, 3–14.

Yeo, K.T. (1993) Systems thinking and project management – time to reunite. *International Journal of Project Management*, **11**, 111–17.

Youker, R. (1992) Managing the international project environment. *International Journal of Project Management*, **10**, 219–26.

Index

in the project management process, 7
objectives, 60, 94
within the brief, 60
Construction management (USA
style), 3, 73
Construction process
model, 220
system, 35
Contingency theory, 37
strategic, 39
Contractor
appointment, 215
integration, 201, 207
matrix structure, 8
relationship with design team, 151,
207
Contracts
standard forms, 11, 143, 151
Control loops, 233, 241
Culture
organisational, 184, 282
traditional, 284

Decisions, 5, 12, 13
feedback, 76, 94, 129
key, 128, 151, 248
on LRA, 236
objectives, 5
operational, 129, 151, 248
on LRA, 236
points, 113, 128
on LRA, 236
primary, 127, 151, 248
on LRA, 236
process, 113
access to, 202
sub-system, 198
Design-and-build, 3, 7, 11, 116, 143,
211
integration, 25, 116, 213
negentrophy, 72
organisation matrix, 207
Design team
integration, 202, 248

Differentiation, 3, 38, 112, 248
definition, 57, 109
determinants, 132
on LRA, 236, 240
in practice, 115

Empowerment, 165, 172
Entropy, 69
Environment, 62, 112
client, 66, 89
definition, 35
differentiation, 38, 73
protected niche, 72
uncertainty, 94, 127
Environmental forces
action of, 66, 248
at start of project, 89
classification, 64
direct, 66
indirect, 66
quantification, 69
Equifinality, 60

Feedback, 75, 248
decision points, 76, 94, 129
managing system, 134, 150
negative, 76
positive, 76
Fees, 24
Firms, 8, 73, 104
objectives, 9
First World War, 20

General Systems Theory, 35
Groupthink, 110

Industrial revolution, 19
Information, 120, 286
Innovation, 10, 21, 38
Input, 91
Institution theory, 40
Integrating mechanism, 97
contingency theory, 37
range, 107